Low Carbon Communities

Low Carbon Communities

Imaginative Approaches to Combating Climate Change Locally

Edited by

Michael Peters

Senior Research Fellow, RESOLVE, University of Surrey, UK

Shane Fudge

Research Fellow, RESOLVE, University of Surrey, UK

Tim Jackson

Director, RESOLVE, University of Surrey, UK

Edward Elgar

Cheltenham, UK • Northampton, MA, USA

Published by
Edward Elgar Publishing Limited
The Lypiatts
15 Lansdown Road
Cheltenham
Glos GL50 2JA
UK

Edward Elgar Publishing, Inc.
William Pratt House
9 Dewey Court
Northampton
Massachusetts 01060
USA

A catalogue record for this book
is available from the British Library

Library of Congress Control Number: 2009942851

Mixed Sources

Product group from well-managed
forests and other controlled sources
www.fsc.org Cert no. SA-COC-1565
© 1996 Forest Stewardship Council

FSC

ISBN 978 1 84844 589 5 (cased)

Printed and bound by MPG Books Group, UK

Contents

Acknowledgements

The editors would like to pay tribute to all of the contributors and reviewers for their genial cooperation, professionalism and attention to detail. We are indebted to Felicity Plester and her colleagues at Edward Elgar for helping to develop the proposal and providing dependable guidance throughout the project. The support of the Economic and Social Research Council (ESRC) is also gratefully acknowledged.

Contributors

Wokje Abrahamse PhD carries out research that focuses on human behaviour in relation to sustainability issues, with a view towards understanding how to encourage behaviour in a more environmentally friendly direction. In particular, she examines the factors that are related to different types of environmentally significant behaviours (for example, energy use, travel mode choice and food consumption), such as attitudes, and environmental values, and the effectiveness of interventions to encourage more sustainable lifestyle choices, and implications for policy.

Elliot Bushay BSc (Hons) Technology, has worked in the field of community engagement and behaviour change in the environmental sector for over seven years, with over five years' experience as a practitioner, applying and developing behaviour change and community engagement principles and techniques. Bushay has a wide range of skills and experience in energy management, environmental monitoring and auditing in the domestic sector. However, Bushay's interests are in individual and community understanding, and relationships with the natural environment, and the learning and behaviour change relevant to achieving sustainable lifestyles.

Mario Cardona holds a Laurea in Adult Education and Training from Roma Tre University, Italy and an MEd in Adult Education from the University of Malta. He is currently reading for a PhD in Adult and Community Education at Edinburgh University. He is a member of the management committee of Koperattiva Rurali Manikata Ltd. His main research interests are social issues, community development, and the pedagogical work of Don Lorenzo Milani and the school of Barbiana.

Ian Christie is an associate of the think-tank Green Alliance and a visiting professor at Surrey University's Centre for Environmental Strategy. He is an independent advisor, writer, teacher and researcher on sustainable development and environmental policy. He is co-author with Michael Carley of *Managing Sustainable Development* (Earthscan, 2000), and with Diane Warburton of *From Here to Sustainability* (Earthscan, 2001). He has worked with many organizations in business, civil society, local governance and central government.

Elizabeth Cox, Head of Connected Economies, New Economics Foundation (NEF) is an economist who leads on NEF's UK and international local economic development work. Her work ranges from campaigns identifying the loss of diversity on UK high streets (Clone Town), through action research to support practical community action in the UK and internationally (BizFizz, Local Alchemy and Plugging the Leaks), to developing approaches to help public bodies to commission and procure more sustainably. New areas of research include developing a low carbon, high well-being economic development model and community action on energy use and generation.

Lara Curran is Senior Policy Officer (Climate Change) at Woking Borough Council. Curran gained a BSc (Hons) in Geography in 2000 and since then has worked in central and then local government. At Woking Borough Council, her key areas of work include monitoring and developing the council's climate change strategy and delivering projects that contribute to its carbon reduction targets.

Scott Davidson MA, MSc is widely involved in the pro-environmental behaviour change field, applying his behaviour change background to make the Global Action Plan (GAP) and other programmes more effective. In doing so he works to act as a bridge between academia and practitioners, and offers research and programme design consultancy to non-GAP behaviour change projects. Davidson's background includes managing international research projects focused on human behaviour and large environmental behaviour change campaigns on behalf of local authorities and NGOs. He has completed an MSc in Sustainable Development and an MA in Psychology.

David Evans PhD is a research fellow at the University of Manchester and an honorary fellow of the ESRC Research Group on Lifestyles, Values and Environment (RESOLVE). A sociologist, his interests lie broadly in consumption, practice and material culture alongside how accounts of these play out when faced with questions of environmental sustainability. Previous projects have focused on marginal space and sustainable lifestyles. Current research interests include food, waste, frugality and the social formation of habits.

Shane Fudge PhD is a research fellow in the ESRC Research Group on Lifestyles, Values and Environment (RESOLVE) and a member of the Centre for Environmental Strategy at the University of Surrey. His research explores energy and environmental regulation from both UK and international perspectives. Recent projects include a critical appraisal of UK energy policy and an internationally scoped piece of research investigating the coordination of an effective synthesis between diverse national energy regimes and the growing influence of global environmental regulation. Both these projects inform the longer term aims of RESOLVE; particularly in developing deeper understandings of the concept of 'governance' and also the wider policy lessons and recommendations which will be used to complement future policy scenario analysis. Fudge holds a PhD from the University of Glamorgan.

Simon Gerrard PhD has more than 20 years' experience in environmental sciences and management. Gerrard has moved from environmental risk management research to practical application. He has been Programme Manager of the Carbon Reduction (CRed) Programme since 2002 and is now the Chief Technical Officer in the newly established Low Carbon Innovation Centre based in the School of Environmental Sciences at the University of East Anglia, Norwich. Focusing on developing, implementing and evaluating carbon management solutions for business, organizations and communities, the Low Carbon Innovation Centre combines essential elements of technological innovation and behavioural change.

Tim Jackson is Professor of Sustainable Development at the University of Surrey and Director of the ESRC Research Group on Lifestyles, Values and Environment (RESOLVE). Jackson has played an advisory role in UK sustainable development policy for more than two decades and since 2004 has been Economics Commissioner on the UK Sustainable Development Commission (SDC). From 2004–2006 he was the sole academic representative on the UK Sustainable Consumption Round Table and co-authored the influential report *I Will If You Will* (SDC, 2006). More recent publications include his controversial and groundbreaking report for the SDC, *Prosperity without Growth* (Earthscan, 2009). In addition to his academic and policy work, Jackson is an award-winning dramatist with numerous radio-writing credits for the BBC.

Victoria Johnson PhD is the lead researcher for the New Economics Foundation (NEF) climate change and energy programme. Johnson has a PhD in Atmospheric Physics from Imperial College, London. Before she joined NEF in January 2007, she worked with a coalition of five local authorities to devise a climate strategy for their region. Johnson is co-founder of onehundredmonths.org, a call for action on climate change, and co-chair of the Roundtable on Climate Change and Poverty in the UK. She recently led a project for Carnegie UK Trust on Bridging the Gap between Social Justice and Climate Change: The Future Role of Civil Society Associations.

Michael Kelly is Prince Philip Professor of Technology at the University of Cambridge, Professorial Fellow at Trinity Hall and a Non-Executive

Director of the Laird plc. He was educated in Wellington, New Zealand and Cambridge. He worked at GEC plc during 1981–92, developing two new families of microwave devices in production with E2V Technologies at Lincoln. He was at the University of Surrey during 1992–2002, including a term as Head of the School of Electronics and Physical Sciences. He was Executive Director of the Cambridge–MIT Institute, 2003–5 and Chief Scientific Adviser to the Department for Communities and Local Government 2006–9.

Howard Lee PhD trained as a plant ecologist and undertook his PhD on the ecology of bracken fern removal from upland pastures. He managed a government potato export breeding programme for ten years and lectured at Queen's University Belfast. Lee then took a lectureship at Wye College (University of London) and taught and managed the MSc programme in sustainable agriculture. At Wye he also researched aspects of sustainable crop production. He is currently at Hadlow College where he has set up, manages and teaches on a degree in Sustainable Land Management. At Hadlow he also works as the Sustainability Champion and was one of the founders of the HadLOW CARBON Community.

Joshua Lockyer PhD is an environmental anthropologist and a post-doctoral teaching fellow in anthropology and environmental studies at Washington University in St Louis, MO, USA. He is a former visiting fellow with The Research Group on Lifestyles, Values and Environment (RESOLVE) at the University of Surrey. He is also on the executive committee of Eco-City usa, a nonprofit organization in St Louis that conducts research on and facilitates the spread of intentional living arrangements. He has been studying ecovillages and sustainability-oriented intentional communities for ten years.

Michael Peters PhD is The Research Group on Lifestyles, Values and Environment's (RESOLVE) Senior Research Fellow and a member of the Centre for Environmental Strategy at the University of Surrey. He is currently coordinating a programme of research on environmental education and community engagement in low carbon social change initiatives. Areas of particular research interest include community-based carbon reduction, sustainable energy, and waste management and environmental policy tools. Peters graduated from Wye College in 1997 with a BSc in Rural Environment Studies and holds a PhD in the Environmental Performance of Small and Medium sized Enterprises from the University of East Anglia, awarded in 2001.

Simon Roberts has been helping people develop effective responses to the threat of climate change and the misery of cold homes since 1985.

Appointed Chief Executive of the Centre for Sustainable Energy (CSE) in 2002, he has previously worked for Triodos Bank and as National Energy Campaigner at Friends of the Earth. Roberts has recently led CSE's work on the distributional impacts of UK climate change policies; establishing new local authority performance frameworks on CO_2 reduction, and community engagement with sustainable energy initiatives. Among his advisory roles, Roberts sits on the Government's Renewables Advisory Board and the ESRC Research Group on Lifestyles, Values and Environment (RESOLVE) programme board.

Gill Seyfang PhD is an expert in low carbon lifestyles, with expertise in researching community-based organizations working towards sustainable consumption. Her model of 'grassroots innovations' offers a new way of examining the potential of local initiatives such as local food systems, complementary currencies and low impact eco-housing, and in particular the scope for such demonstration projects to scale up and influence mainstream consumption patterns. Seyfang is the author of *The New Economics of Sustainable Consumption: Seeds of Change* (Palgrave Macmillan, 2009).

Justin Spinney PhD is a research fellow in The Research Group on Lifestyles, Values and Environment (RESOLVE) and a member of the Centre for Environmental Strategy at the University of Surrey. His research seeks to understand the processes which lead to the normalization and escalation of everyday practices surrounding the use of specific personal computing appliances. Research interests include the relations between instrumental, embodied and symbolic aspects of consumption; the role of space and place in (re)producing values and behaviours; the use of video within ethnographic methodologies and particularly in relation to mobile and embodied practices; the construction and negotiation of environmental meanings within businesses and stakeholder organizations; and sustainable mobilities.

Julie Taylor holds a BA Hons degree in Three Dimensional Design (Theatre Design) from the Birmingham College of Art. She has worked at Harker Brothers, London (Scenic Artists), painting scenery for Covent Garden Opera House and Glyndebourne Opera House. She initiated and curated the Finite Environmental Art Exhibition at Hadlow College. Taylor now works as an artist, exhibiting locally, in London and the US, and as a part-time teacher.

Preface: The Research Group on Lifestyles, Values and Environment (RESOLVE)

Funded by the Economic and Social Research Council as part of the Research Councils' Energy Programme, RESOLVE is a unique interdisciplinary collaboration spanning four internationally acclaimed departments at the University of Surrey: the Centre for Environmental Strategy and the Departments of Economics, Psychology and Sociology. RESOLVE's research programme is arranged around five thematic research strands: Carbon Mapping; Psychology of Energy Behaviours; Sociology of Lifestyles; Carbon Lifestyle Scenarios; and Governance for Sustainable Lives. Recent findings to have emerged from RESOLVE's interdisciplinary research programme are reported among the chapters in this volume.

The guiding ethos for our work is to combine academic excellence with policy relevance. An explicit goal is to provide robust, evidence-based advice to policy makers who are seeking to understand and to influence the behaviours and practices of 'energy consumers'.

RESOLVE hosted a successful international conference at the University of Surrey in December 2008, 'Community Action on Climate Change', where academics, practitioners and policy makers came together to explore the social drivers and barriers to effective community engagement in carbon reduction. This volume complements those discussions and moves the debate forward.

Foreword

Ian Christie

At the time of the Rio de Janeiro Summit on Environment and Development in 1992, the then US President George Bush Senior declared that whatever else was on the agenda, the American lifestyle was not up for negotiation. Although few other heads of state and political leaders in the rich world have made such uncompromising remarks about the untouchability of their domestic mode of consumerism, there has been huge reluctance among politicians and policy makers to contemplate 'lifestyle change' and any notion of limits to consumption in devising strategies to meet the enormous challenges posed by the risk of climate disruption, ecological breakdowns and depletion of key resources.

The reasons for this are plain enough. First, the economy and culture of mass consumer affluence have been established in the West for half a century or more, and have been felt to 'deliver the goods' with unprecedented success. The idea that this enormously productive system might run up against ecological limits is deeply unwelcome, and one that is bound to be resisted by many. Policy makers and many citizens hold to a picture of the world, a seemingly robust one, in which new technologies will always come to the rescue, as they have flowed in profusion throughout living memory.

Secondly, democratic political competition has been geared for decades to the idea that living standards and consumption can continue to rise indefinitely. There is no mainstream vehicle or ideology for articulation of any alternative 'frame' or paradigm for economic and social life. Thirdly, the UK and many parts of the West are now a thoroughly liberalized set of polities and economies. The 1960s' cultural revolutions have led to a widespread, if still contested, pattern of social liberalism; the 1980s' economic revival of liberalism led to a similarly widespread and still more contested deregulatory paradigm in policy making and governance. In this world, individual choice in lifestyle is supreme, subject only to restraints established in law. At worst, this has led to a hollowing-out of the very idea of the public good and of common values and standards that trump individual choices in the marketplace. It has also combined with other 'atomizing' forces in society to weaken the institutions, civil associations and everyday practices of cooperation at local level between the state and the individual. That too makes for a barrier to political action: if environmental policy pressures demand 'lifestyle change', then there seems to be little else to do other than set up a one-sided process of exhortation between state and individuals, typically via information campaigns that have little effect because they exist in a social vacuum.

This is an unpromising context for promoting changes in behaviour on a societal scale. But the nature of the ecological challenges at hand tells us that precisely this is what is needed. Why? To see the fundamental reason, we need to consider the famous 'Sustainability Equation' published in 1974 by Paul Ehrlich and John Holdren, $I=PAT$. Here I is ecological impact of development, which in turn is the product of three factors – *P*opulation, *A*ffluence (i.e. levels of consumption) and *T*echnology (efficiency of production and consumption). We face a rising global population for sure; rising consumption demands for sure; and possible gains in technology that can boost production of key resources (food) and improve efficiencies (energy and material reduction, recycling and reuse). The rise in population and consumption that is in store to the middle of the century is so great that in order to reduce our already unsustainable collective impact on the Earth's climate system, resources, sinks and biodiversity, a near-incredible continuous improvement in resource efficiency would need to take place in coming decades. (For more on this, see Tim Jackson's masterly *Prosperity without Growth*, Earthscan, 2009.) Now, it may be that such gains can be made and that 'rescue technologies' will arrive. But no-one can count on this: the lesson of recent decades is that advances in marginal efficiency of key technologies (vehicle engines, for instance) are overwhelmed by the scale of increase in consumption. There is no reason to suppose this problem will disappear; and every reason to suppose it will get worse as population grows and with it the justifiable demands of the global poor for better lives and more consumption.

Technological innovation is crucial to the transition to a low carbon economy and sustainable patterns of development, but while necessary it is far from sufficient. Population growth has slowed and global population may well stabilize in the middle of the century, but we want that to happen through peaceful and gradual social change, not to be precipitated by ecological disruption. So we are left with the third term in the Sustainability Equation – consumption – when it comes to making sure of a successful transition from unsustainable and fossil-fuel-dependent societies. Large parts of the world need to be able to consume more, for reasons of equity, justice and poverty reduction; so the rich world needs to consume differently, and with far less impact, to allow room for the poor to raise their quality of life and to allow for ecological recovery. This means consuming some things differently; consuming some new services and products; substituting some for others; and in some respects – such as aviation and car use – doing with less than we have become accustomed to. All that adds up to 'behaviour change' – and to add to the complexity, it must be negotiated and designed to ensure that social inequalities are not worsened. The moment limits to consumption are invoked, attention falls on the existing distribution of opportunity and resource use. Lifestyle change is a distributional issue as well as a problem of motivating changes in behaviour, attitudes and values.

This all helps explain the reluctance of politicians and policy makers to speak plainly about the inescapable need for changes in consumption, including real reductions in some carbon-intensive activities. However, the weight of evidence about actual and possible climate change, and the complexity and time-lags involved in radical changes to the supply side of energy systems, mean that this reluctance has been diminished. There is no choice but to confront lifestyle change and the question of how to influence behaviour, attitudes and values.

What has become apparent is that systematic knowledge is lacking: policy making has been based on hunches and rough working assumptions about behaviour, or on deeply flawed models of agency such as the individualist rational choice paradigm in economics. So there is now a small but vital 'industry' working on behaviour change issues in relation to environment and sustainability across Whitehall, in public agencies, in local government, all hungry for robust evidence and guidance from academic research.

The University of Surrey's RESOLVE programme on lifestyles, consumption and low carbon living is at the leading edge of academic research, with many emerging results that are highly relevant to policy makers. This volume provides many illustrations of the valuable insights emerging from RESOLVE and other programmes and projects in the UK and beyond. One of the major themes of social scientific research in this field concerns the implications of the fundamental insight that a great deal of what appears to be autonomous 'individual choice' is *socially* mediated and 'constructed'. Our desires are our own, but they are shaped and influenced by social context – what our peers deem desirable, what role models and other influencers are doing and saying, what social norms tell us is acceptable, what we feel we need to want and be like in order to be accepted by others and make a respectable place for ourselves in our social worlds. A key implication, supported by many findings, is that when we change our behaviour we do it as a result of, and in tandem with, wider socially validated changes. It is far easier for an individual to change consumption choices and behaviour if the surrounding social environment helps her to do so, validates the choice, does not make her feel as if she is 'standing out', and enables her to make changes along with a suitably large and convincing body of peers. In other words, individual change is made much more likely and easy if it is part of a collective shift. Social incentives help make new individual behaviour possible.

This is important because it helps us understand why the prevailing policy model for inducing behaviour change is so weak. Governments and agencies have often assumed that provision of information will lead to individuals rethinking what they do, and making a change. In practice, information may be necessary but is invariably insufficient. People need to make changes to lifestyles via shifts in social norms and incentives that help move whole sections of society. This leads us to focus afresh on old and new forms of association and social interaction lying at levels between the state and the individual, for it is here that the influence of peers, crowds and role models may be most powerful. It is also here that old lessons about the importance of small-scale, 'intermediate' civil associations in social life and changes in norms and behaviour may need to be relearned and applied in new settings. For the social and economic revolutions of the 1960s and 1980s have been damaging to many such institutions (political parties, trade unions, churches, voluntary bodies, neighbourhood organizations and informal local voluntary arrangements between households).

A great deal of effort is now being made to think through how we might reinvigorate these intermediate organizations of civic cooperation and innovate with existing ones and new forms of association. The widespread political move to rethink 'localism' has been born of disillusionment with central bureaucracies and excessive control over local governance, but it could have a positive dimension in relation to sustainability. Central government is neither trusted nor uniquely capable when it comes to promoting behaviour change for sustainable living, but there could be a renewed role for local government and civil institutions at local level in this field. Local councils are closer to the citizen and are certainly more efficient at 'delivery' than much of central government has been. And there is a hunger for new forms of 'community spirit' and cooperation, as a reaction against excessive and atomising individualism and 'marketization' over the past generation, and as a key part of the search by many people for ways to reduce their impact on the environment and contribute to sustainable living for all. This brings together innovation in old forms of association – such as projects led by many people in faith communities, increasingly concerned with environmental action; and innovation in new vehicles for cooperation and collective behaviour change, such as the fast-growing Transition Town movement.

These are exciting developments, illuminated and assisted by leading academic research and thinking of the kind on display in this volume. Few aspects of the transition to sustainability are more important than the search for effective means of enabling behaviour change, innovation and cooperation in local communities. This collection is a hugely valuable contribution to our understanding, and to the work of practitioners and policy makers alike.

Introduction

Michael Peters, Shane Fudge and Tim Jackson

The climate challenge is above all a challenge for governance. Fossil fuels have been for two centuries or more the invisible currency of everyday life. They are deeply implicated in the complex patterns of modern living. How we live, how we work, how we eat, how we get around: all depend intimately on access to high quality energy sources. This means primarily a dependency on the consumption of fossil fuels. And that in its turn has led to carbon emissions at levels which are now clearly recognized as unsustainable. Shifting those patterns, weaning ourselves from carbon dependency, is more than just a technological task. It calls on society to untangle the structures of provision and build new, more sustainable practices, infrastructures and forms of social organization. Governance is key to this endeavour.

The challenge of climate change for governance is, in part, a challenge about scale. 'Think global, act local' emerged in the 1990s as a familiar axiom in response to global environmental challenges like climate change. The new millennium brought a dawning realization of how difficult this exhortation was to follow. Global challenges demand global responses. International climate policy claimed centre stage in environmental politics through the Kyoto Protocol. And yet achieving the goals of the Protocol proved elusive. A part of the reason for this was the failure of Kyoto to create a global emissions cap. But even the playing out of reduction targets to the national level turned out to be no recipe for success.

The limitations of top–down national governance in addressing the urgency of climate change are vividly illustrated in the UK during the 1990s (Harding and Newby, 1999). Energy privatization had resulted in a 'dash for gas' and a contraction of the coal industry. As a result the UK's greenhouse gas emissions appeared to decline during a period in which climate change started to become a growing international concern. This structural shift enabled the Labour Government of 1997 to set out an early target of a 20 per cent reduction in the UK's CO_2 levels by 2010 according to 1990 baseline levels – more demanding even than the Kyoto target.

But the rather more complex nature of the issue soon became apparent when emissions began to rise again during this time, mostly due to continuing growth in road transport and air travel (Royal Commission, 2000). There was also mounting evidence to suggest that there would now have to be a more direct connection in policy initiatives to household energy demand – responsible for as much as 40 per cent of the UK's CO_2 emissions total (Jones et al., 2000). Critics pointed out that this was evidence that policy initiatives needed to be much more flexible if they were to be effective in addressing climate change.

The reality is that governance itself becomes 'stretched' by the demands of climate change. It must reach upwards to the world stage; downwards to regions, local communities and households. Global targets must mean something to households. Global initiatives must resonate at the local level. Communities must play a crucial part in the protection of the global commons.

In the UK, this dynamic has begun to evolve into a complex, multi-level political structure; demonstrating many of the hallmarks of what Hooghe and Marks (2001) have described as a system of 'multi-level governance'. The UK's climate change policy framework exemplifies this shift. Decision-making and implementation are coordinated through a complex network of intersections within and between national, international and local levels. The loci for decision-making have become dispersed across a variety of institutional structures.

The importance of the local level in this complex 'political triad' has become particularly significant in recent times, where a focus on the behavioural and social changes which will be needed to adapt human society to the constraints of the planet has become increasingly apparent. This recognition has increasingly informed government policy and academic inquiry during the last decade (Jackson, 2005, 2008), and has provided the impetus for a diverse range of emergent 'grassroots' initiatives (Church and Elster, 2002; Jackson and Michaelis, 2003).

Perhaps surprisingly, this emphasis on local initiatives for social change draws strength from a long intellectual pedigree. At its broadest level, casting the climate action in terms of behavioural change suggests a particular manifestation of a perennial social issue. As Gardner and Stern (2002) have pointed out, it is essentially the problem of ensuring that behaviours which threaten the well-being of the social group are discouraged and that those which promote long-term well-being are encouraged. In other words it is quite precisely the problem of coordinating individual behaviour for the common good.

Ophuls (1973) argued that, from time immemorial, there have only ever been a few basic methods – written about by philosophers and employed by societies – for achieving this. Specifically, the four 'solution types' are (Gardner and Stern, 2002, p. 27):

1. government laws, regulations and incentives;
2. programmes of education to change people's attitudes;
3. small group/community management; and
4. moral, religious and/or ethical appeals.

For the most part, modern governance has tended to favour the first two strategies, although there is evidence of a returning allegiance to moral exhortation (for instance in the UK Government's *Act on CO₂* campaign). Gardner and Stern (2002) call community management the 'forgotten strategy' because of its prevalence in early forms of social organization and relative absence in modernity. There are of course some clear reasons for this shift, not the least of which is that community itself has become a casualty of increasingly globalized economies (Putnam, 2000; Jackson, 2009).

Nonetheless, in recent years the UK Government has been particularly keen to emphasize the important function that local government and communities can – and should – perform in galvanizing action towards household carbon reduction. The recent publication of the *Low Carbon Transition Plan* (DECC, 2009) confirms the importance of this agenda. A whole chapter of this latest energy White Paper is devoted to the significance of 'transforming our homes and communities' as part of an integrated effort towards meeting the UK's legally binding target of at least an 80 per cent reduction in greenhouse gas emissions below 1990 levels by 2050. Prior to this, many UK local authorities had already begun the process of both encouraging and introducing community-oriented projects, focusing on changes that individuals and households themselves could make in order to save energy and reduce carbon emissions. During the same period, several 'non-state' community projects responded independently to the pressing urgency of directing society towards more sustainable ways of living, again keeping a focus on the important contribution that groups of people – and alternative social and economic arrangements – can make towards enabling a lower carbon future connected with more sustainable lifestyles.

Again the rationale for these initiatives is well established. What makes community management systems work, according to Gardner and Stern (2002), is a combination of participatory decision-making, monitoring, social norms and community sanctions. Interestingly, sanctions and penalties for non-compliance are not the most important element in ensuring compliance. Rather, the effectiveness of group management comes from the internalization of the group's interest by individuals in the group. Gardner and Stern suggest that there are several reasons why people internalize group norms. In the first place, they have participated in creating them. In the second place, they can see the value of these norms for themselves in preserving and protecting the interests of the local community and themselves as members of that community.

In addition, these group norms become a part of the shared meaning of the community and contribute to the social well-being of the group, not just through the protection of resources but through the development of trust, collaboration and social cohesion. Sanctions may be necessary to protect the group from those tempted to violate the collective good for individual interests, but the main reason people accept and act on social norms is that doing so cements social relations, signals membership of the group and contributes to a sense of shared meaning in their lives (Jackson, 2005).

This is not to suggest of course that community-based strategies are without drawbacks. In fact, as some of the contributions in this volume highlight, even the matter of defining and working with communities is far from straightforward. It is clear that society is decreasingly made up of integrated, geographically based communities in a way that it once was (Bauman, 2002) and is increasingly made up of numerous diverse communities – communities which often overlap and sometimes exist in complete isolation from one another.

In today's increasingly technically oriented system of networked communication, one person may well engage in a variety of different communities simultaneously. In the UK this has taken on further complexity due to the shifting role of local government over the last two decades, encouraged by political devolution. The relevance of all these issues in terms of coherent strategies for action and effective engagement means that there is clearly no one approach or project which is necessarily going to reach 'the community' and, by extension, the issues within it. As policy makers and practitioners are now acknowledging, a diversity of tools and approaches is required.

Here then lies both the opportunity and the challenge of community-based action on climate change. On the one hand, it offers a meaningful way to achieve global targets and an avenue for renewing social relations at the community level. On the other, it challenges fundamental aspects of social organization in the modern economy and sometimes runs foul of wider structures and constraints. These are the issues which this book sets out to explore.

The first section of the book provides a series of theoretical and intellectual understandings of communities and social change supported by insights from a range of empirical research studies. In Chapter 1, Peters provides an introductory overview of how the term 'community' has been conceptualized in sociological literatures, noting that there remains considerable uncertainty with regard to the way in which communities could or should be defined. The chapter examines the salience of underlying concepts of social organization that can shape and influence the extent to which programmes of engagement are likely to be successful. Drawing on recent empirical work some of the key opportunities and challenges for local government in translating the concepts into practice are considered.

The inter-play between individuals takes on particular significance when considering the role of 'intentional communities'. Evans (Chapter 2) takes a fresh look at the processes that lie behind the evolution of values and how these play out in community life with regard to people who choose to identify themselves as living sustainable lifestyles. His ethnographic fieldwork points to the importance of moral regulation; a property of the community concept that goes beyond emotional attachment and identification with others. It is suggested that individuals who are attempting to live sustainably move between attachments to various collectives and a sense of detachment in order to pursue the multiple agendas that a (sustainable) lifestyle demands in the modern world.

It is often pointed out that a clear shortcoming of many community-based programmes of engagement and intervention (both state and non-state led) has been their failure to establish, from the outset, rigorous monitoring processes. As Abrahamse (Chapter 3) points out, this is crucial for the proper evaluation of such efforts. Assessing the role of group membership, social norms, social identities, threatened identities and social comparison processes as potential motivators for change, this chapter provides an overview of theoretical approaches to social change. This is complemented with a critical review of several intervention studies that have tried to make use of group dynamics in order to encourage change towards a lower carbon future.

In Part II, attention is turned to challenges for local level climate policy in supporting the transition to low carbon economies. Consideration is given to alternative models of local level governance, decision-making and economic frameworks. In Chapter 4 (Fudge) the development of environmental policy in the UK from the post-war period is examined, drawing attention to the substantial changes that have occurred over the last fifty years and the range of influencing factors that have impacted upon and modified the formation of nation-state politics in this period. The chapter considers how national policy agendas are increasingly contextualized in relation to European legislation, international agreements and local ascendancy. The targeting of individuals and behavioural change necessitates local government and community-led action as a central facet of the political process.

This theme is picked up by Roberts (Chapter 5), who provides an investigation of the potential extent of local authority influence in galvanizing community action on climate change with particular emphasis on key issues for practitioners. The chapter explores the circumstances under which effective engagement and action might be realized and the comparative advantage associated with a variety of approaches being used in attempts to establish these conditions.

The focus for Chapter 6 (Spinney) is on partnership working and green alliances, highlighting some of the tensions encountered between non-governmental organizations and local government in operationalizing cycling as a sustainable mode of urban mobility. Particular problems associated with working in a low carbon policy community are highlighted, noting that certain stakeholders are in fact excluded and transformed through the multi-faceted process of partnership working.

The transformation of lifestyles in pursuit of a sustainable society connects closely with intellectual investigation in the field of sustainable consumption. In Chapter 7, Seyfang presents an holistic 'New Economics' agenda for sustainable consumption which involves new ways of working, redefining the measurement of value and progress, new uses for money and expressing ecological citizenship. Applying these ideas to practice, the chapter presents examples of complementary currencies (time banks and local money systems) and critically assesses their potential as tools for carbon reduction and sustainable development.

The question of whether current forms of organizational structure and governance are capable of supporting and encouraging more sustainable lifestyles at the regional and sub-regional level is explored by Cox and Johnson (Chapter 8). Here, attention is drawn to the inter-connectedness of the economy, environmental constraints, societal values and global interdependence. It is argued that a first and necessary step to decarbonizing local economies lies in the recognition that it is possible to change and re-engineer the economy to better serve societal goals at the national and local level. The chapter sets out a novel range of propositions for an effective approach in support of communities making the transition to low carbon, 'high well-being' lifestyles.

The final section of the book, Part III, incorporates a variety of models and case studies, showcasing the opportunities and challenges for community-focused action in a diverse range of settings and contexts. The first example given (Gerrard, Chapter 9) is the Community Carbon Reduction Project (CRed) – one of the more established carbon reduction programmes in the UK. The project tracks and monitors progress of participants along carbon reduction pathways related to home and work life, encouraging them to do more. CRed's ethos is that small things do add up and that 'we are all in this together'. Periodic evaluation seeks to establish the reality of community carbon-saving, providing important evidence into the progress made by individuals as a contribution to meeting the challenges of climate change.

Davidson (Chapter 10) outlines Global Action Plan's (GAP's) EcoTeams programme, including evidence gathered from participant feedback and measured savings. The programme is assessed with reference to current behaviour change literature, highlighting the underlying basis of GAP's methods and drawing attention to implications for the wider field and the role of government.

The sort of positive progress that can be realized through a pragmatic climate change strategy in local government and effective communication of climate change messages to the public is illustrated by the example of Woking Borough Council. In Chapter 11 (Curran) Woking's approaches are considered, exploring a range of techniques and initiatives pursued and established by the Council, including the formation and activities of its energy and environmental services company Thameswey Ltd.

Ecovillages and transition towns are examples of local community initiatives for carbon reduction and climate change action that continue to make progress without specific prompting or assistance from government agencies or policy. Lockyer (Chapter 12) describes the types of activities being undertaken in these communities and considers their effectiveness within the context of the challenges posed by climate change. The author also explores participants' motivations for engaging in these initiatives and investigates the policy obstacles and opportunities that they have faced along the way.

In Chapter 13, Bushay describes the Energy Conscious Households in Action (ECHO Action) project – a novel 2 and-a-half-year initiative which ran from 2006 to 2009 and was part-funded through the European Union under the Intelligent Energy Europe Programme. Its underlying aims and approach are set out and findings to have emerged from two London-based groups and other European partners presented. The chapter articulates how progress made by ECHO Action fits into the context of similar models of community engagement for improved household energy and environmental performance, including an assessment of participant profiles. Attention is drawn to the problem of broader community engagement beyond enthusiastic, already 'pro-environmental' individuals.

The issues of sustainability and quality of life in rural communities are addressed in Chapter 14 (Lee and Taylor), which takes a close look at the activities of a recently established low carbon community initiative in the

village of Hadlow, Kent. A unique feature of this case study is the collaboration between members of the village community and the local college
of further and higher education (Hadlow College), where the project's
inception and development began. The chapter demonstrates how self-
motivated communities can conflate reduced fossil carbon emissions with
access to sustainable supplies of energy, food and water. The process of
empowering project participants to embrace positive action has facilitated
a range of additional benefits including improved mental and physical
well-being.

Another, quite different, rural community case study is presented in
Chapter 15 (Cardona). Here the author describes how the reaction of a
small farming community in Malta to an imposed radical land-use change
by the Maltese Government has resulted in a socially cohesive drive
towards new opportunities for sustainable agriculture and locally sourced
products. The restoration of previously inaccessible archaeological and
historical buildings and monuments has been integrated into the project
with a view to encouraging the general public to appreciate the natural
beauty of the surroundings as well as agricultural life in the area.

In summary, the contributions to this book highlight both the promise
of community-based action on climate change and some of its limitations.
It is clear immediately that collective action at the local, community level
provides many opportunities for contributing to and enabling the transition to a low carbon economy. This potential is not always straightforward
– partly because of the eclectic nature of community configurations and
partly because of the existence of wider constraints on community action.
But it remains worth pursuing as an integral facet of a more cohesive drive
towards sustainability.

When it comes to overcoming those wider constraints, there is a clear
need to think 'outside the box' and consider the potential for alternative
and imaginative models and approaches to connecting with individuals.
For example, exciting new possibilities emerge around alternative currencies and the re-engineering of local economies.

Although many local authorities across the UK have been slow to take
up the challenge of community engagement in practical ways, the evidence
in this book suggests that they have an increasingly important and influential role to play, in their capacity as an interface between citizens and
government policy. An emerging – and growing – body of evidence demonstrates that the relatively small numbers of 'best practice' cases are in
fact making some noteworthy progress that might usefully be replicated,
with suitable modifications, on a broader scale.

In the final analysis, communities come in many forms – a mixture of
policy tools and practical 'handles' which people can identify with and

latch on to will be essential. Community action is about collective endeavour towards a common goal – a better future. It is the antithesis of isolation and individualism.

REFERENCES

Bauman, Z. (2002), *The Individualized Society*, Malden, MA: Blackwell.

Church, C. and J. Elster (2002), *Thinking Locally, Acting Nationally: Lessons for National Policy from Work on Local Sustainability*, York: The Joseph Rowntree Foundation, accessed at www.jrf.org.uk.

DECC (2009), *The UK Low Carbon Transition Plan: National Strategy for Climate and Energy*, London: Department for Energy and Climate Change, and Norwich: The Stationery Office, accessed at www.decc.gov.uk/en/content/cms/publications/lc_trans_plan/lc_trans_plan.aspx.

Gardner, G. and P. Stern (2002), *Environmental Problems and Human Behaviour*, 2nd edn, Boston, MA: Pearson.

Harding, L. and L. Newby (1999), 'Local Agenda 21 – cutting the economic mustard', *EG-Local Environment News*, **5** (7), 2–3.

Hooghe, L. and G. Marks (2001), *Multi-Level Governance and European Integration*, Lanham, MD: Rowman and Littlefield.

Jackson, T. (2005), *Motivating Sustainable Consumption – a Review of Evidence on Consumer Behaviour and Behavioural Change*, London: SDRN.

Jackson, T. (2008), 'The challenge of sustainable lifestyles' in G. Gardner and T. Prugh (eds), *State of the World 2008 – Innovations for a Sustainable Economy*, Washington, DC: Worldwatch Institute, Chapter 4.

Jackson, T. (2009), *Prosperity without Growth – Economics for a Finite Planet*, London: Earthscan.

Jackson, T. and L. Michaelis (2003), *Policies for Sustainable Consumption*, London: Sustainable Development Commission.

Jones, E., M. Leach and J. Wade (2000), 'Local policies for DSM: UK's Home Energy Conservation Act', *Energy Policy*, **28**, 201–11.

Ophuls, W. (1973), 'Leviathan or oblivion?', in H. Daly (ed.), *Towards a Steady State Economy*, San Francisco: W.H. Freeman and Co, Chapter 8.

Putnam, R. (2000), *Bowling Alone: The Collapse and Revival of American Community*, New York: Simon and Schuster.

Royal Commission (2000), *Report on Environmental Pollution*, London: The Royal Commission on Environmental Pollution, accessed at www.rcep.org.uk.

PART I

Facilitating the low carbon transition:
theoretical and intellectual understandings of
communities and social change

1. Community engagement and social organization: introducing concepts, policy and practical applications

Michael Peters

INTRODUCTION

Community efforts to cut carbon dioxide (CO_2) emissions hold the potential, in principle, to benefit from a greater emphasis on collective expediency; that is, accomplishing more by acting together rather than alone, with broader motivational impacts in terms of encouraging positive beliefs and actions (Walker and Devine-Wright, 2008; DECC, 2009). The involvement of households in this drive needs to be a core feature of attempts by policy makers seeking to address climate change locally and nationally. In the UK, for example, homes are responsible for 27 per cent of total CO_2 emissions nationally (House of Commons, 2009) and the energy used by households for water and space heating accounts for 13 per cent of the UK's total greenhouse gas emissions (DECC, 2009).

For some time now there has been a growing consensus amongst policy makers that projects rooted in bottom-up social, cultural and economic arrangements hold the potential to be more effective than top-down solutions by a) enabling individuals to recognize their own role in contributing to more sustainable energy use and b) providing greater encouragement to citizens to engage more fully in the wider political debate on sustainable living (Long, 1998; Jordan, 2006; Fudge and Peters, 2009). The role of local government in tackling climate change in the UK (including the stimulation of individual and collective household action) is increasingly identified as a key component of a concerted national effort to curb carbon emissions (CSE, 2005, 2007). Both the 2007 White Paper *Meeting the Energy Challenge* (DTI, 2007) and the 2009 White Paper *UK Low Carbon Transition Plan* (DECC, 2009), for example, emphasize the important role to be played by the local government sector in fostering this agenda, and a set of new performance indicators has been adopted such that local authorities are now measured on how well they use their influence to cut back emissions locally (CLG, 2007a).

Several local authorities in the UK have been proactive in these regards and have driven forward the establishment of various community engagement programmes and projects designed to connect their residents into processes of social change and more sustainable ways of living (Fudge and Peters, 2009). While some projects have been structured around the collective adoption of a new technology (for example, the installation of a renewable energy plant) others have focused more directly on providing educational signals and support mechanisms in order to encourage the improvement of energy management in and around the home.

This chapter begins by considering the notion of 'community' through an exploration of some characteristic concepts by which academics and policy makers have tried to make sense of the term, that is social capital, social learning and social norms. It will be argued that it is important, particularly at a local policy level, to have clarity on these concepts of social organization in order to inform appropriate designs for effective community engagement programmes. The second section introduces themes that will be revisited in Chapter 5 (Roberts), examining the increasing focus on the role for UK local government in mobilizing social change towards a lower carbon future. The characteristic elements of this approach to the governance of climate change are then illustrated and explored in more detail through empirical evidence gathered during interviews with project officers from three English local authorities. Some of the difficulties for local government in translating key concepts of positive social organization through to practice are highlighted, in particular the link between these and broader social challenges of trust and public perception.

COMMUNITIES AND SOCIAL ORGANISATION

A critical challenge for implementing programmes of prevention and intervention within communities revolves around how to arrive at an appropriate definition of 'community'. It has been argued that the term itself involves two related ideas: first that the members of a group organized around community have something in common with each other; and secondly that the thing held in common distinguishes them in a significant way from the members of other possible groups (Cohen, 1985). In this sense 'community' is a relational concept that implies both similarity and difference and, as Cohen argues, is generally used to establish what it is that distinguishes various social groups and entities.

While the concept of 'community' has been an enduring and recurring theme in modern social science, it nevertheless remains elusive and somewhat intractable with regard to specific definition and meaning (Cohen, 1985; Crow and Allan, 1994; Delanty, 2003). Newby (1994) argues that communities are intrinsically linked to deep-seated emotions, sentiments and beliefs. In these regards, 'common sense' understandings of what community means and represents often refer to readily accessible concepts such as nationality, location, knowledge of location, neighbourhood and language (Studdert, 2006). This type of definition has been synthesized into two broad categories: 'communities of place' and 'communities of interest' (Pelling and High, 2005). The former concentrates on people within a defined geographical area (for example, a particular neighbourhood, or a housing estate) while the latter (also called 'interest groups') focuses on people who share a particular experience, demographic characteristic or interest (for example, the working population, young people, disabled people, ethnic groups, etc.).

These various levels and sizes of community are linked temporally, spatially, physically and psychologically in a wide range of ways, from a global community scale down to very small groups of individuals. This would suggest that the communities people belong to are, as such, multiple and overlapping. Studdert (2006) reasons that this communal multiplicity is not something that people have a great deal of control over, but is rather a product of the social interactions that each individual inhabits and is born into. Similarly, from a theoretical perspective, approaches to conceptualizing and understanding community tend to be overlapping and entwined. For example, community can be viewed both as a value and as a descriptive category, or a set of variables (Frazer, 1999). Community relies on the presence of other people and on action and speech; in essence 'having something in common' (Willmott, 1989).

Several writers and studies have attempted to reveal the extent to which communities as social groups are able to act as conduits for adopting better (in terms of social justice and environmental sustainability) or more desirable patterns of thought and behaviour (see for example Chaskin et al., 2001; Sampson, 2002; Mancini et al., 2005). Consideration of some of the conceptual arguments underlying the role of communities in enabling change and sharing best practice can be useful in informing community-based behavioural change programmes and initiatives. In the following sections three key concepts are considered in this way: social capital, social learning and social norms.

Social Capital

Social capital refers to features of social organization that augment a community's (and, more broadly, society's) productive potential (Putnam,

1993; Roseland, 2000). The main, broad areas for consensus within the social sciences regarding the definition of social capital centre on three core components: social networks, social norms and sanctions (the processes that help to ensure that network members keep to the rules) (Healy, 2001; Halpern et al., 2002).

Three main, distinguishable 'types' of social capital have been identified in the literature (Halpern et al., 2002):

- 'bonding' social capital (for example, among family members or ethnic groups), characterized by strong, often emotional bonds;
- 'bridging' social capital (for example, ties across ethnic groups), characterized by weaker, less dense but more cross-cutting ties; and
- 'linking' social capital (for example, links between different social classes), characterized by connections between those with differing levels of power or social status.

Because social capital is largely developed independently of the state or large corporations (for example, in terms of the organizations, structures and social relations that people build up themselves), Putnam (2000) argues that it provides an important function in terms of strengthening what he calls 'community fabric'. High social capital, in the form of social trust and associational networks, has been signposted by the literature as being linked to a wide range of desirable policy outcomes (Halpern, 2001). For example, Putnam (2000) points to the powerful and quantifiable effects of social capital on many different aspects of our lives, asserting that the concept is more than 'warm, cuddly feelings or frissons of community pride' (p. 279). Various authors have taken this idea further in identifying a range of such quantifiable effects (ONS, 2001), which include the following:

- lower crime rates (Halpern, 1999; Putnam, 2000);
- better health (Wilkinson, 1996);
- improved longevity (Putnam, 2000);
- better educational achievement (Coleman, 1988);
- greater levels of income equality (Wilkinson, 1996; Kawachi et al., 1997);
- improved child welfare and lower rates of child abuse (Cote and Healy, 2001);
- less corrupt and more effective government (Putnam, 1995).

One uniting hypothesis stemming from this evidence appears to be 'receptivity' – that is, the higher the level of social capital the more likely a community is to respond positively to influences that attempt to encourage change for the better; socially, economically and environmentally. The implication of a strong stock of social capital is potentially relevant, therefore, when considering the role that communities might play in moving society forward towards more sustainable ways of living.

Social capital, like other forms of capital, can be productive, but if it is not renewed it can be depleted (Coleman, 1988). Advocates argue that more social capital will be produced the more people work together but, conversely, community stocks of social capital will deplete the less people work together (Cooper et al., 1999; Putnam, 2000). It is thus important to devise and enable strategies for its maintenance and renewal, thus protecting the 'social ozone' (Healy, 2001).

Persuasion and Social Learning

At its simplest level, the premise for wanting to enact change in a community away from unnecessary wastefulness of resources and towards more sustainable patterns of living is because a path characterized by the former is likely to threaten the well-being of both the environment and the social group itself. In terms of governance, the problem has been seen as being one of providing individuals within a group with the right mix of incentives and disincentives as a means through which to coordinate individual behaviour for the common good (Jackson, 2004). Gardner and Stern (1996) argue that community management is more likely to be effective in groups where there are *existing* networks and widely shared norms before the problem of resource management becomes prominent.

Information and Persuasion

Utilizing the strengths of existing networks of communication can be one of the most effective means for disseminating *information*. The ways in which such 'information' or social signals are processed by and become influential upon individuals have been scrutinized in the fields of persuasion theory and social learning. The Hovland–Yale model of persuasion, for example, incorporates three central features of potential influence, namely 1) the credibility of the speaker (the source), 2) the persuasiveness of the argument (the message) and 3) the responsiveness of the audience (the recipient) (Hovland et al., 1953; Hovland, 1957).

Jackson (2005) highlights the limitations that this model has attracted from critics including Petty et al. (2002) and Greenwald (1968); especially regarding the implicit assumption that attitude change occurs through the assimilation and comprehension of persuasive information and that this automatically leads on to shifts and changes in behaviour. As Jackson argues, in reality, 'empirical evidence indicates that learning can occur without any change in attitudes and that attitude (and behaviour) change can occur without any assimilation of the persuasion message' (Jackson, 2005, p. 96).

In terms of developing strategies designed to engage individuals and their communities in processes of attitudinal and behavioural change, levels of existing – or latent – motivation are critical; as is the ability of individuals to make those changes. Petty and Cacioppo have incorporated this recognition into a more recent model of persuasion (the 'elaboration likelihood' model), which distinguishes between central and peripheral processing; two distinct types of psychological processes involved in attitudinal change (Petty, 1977; Petty and Cacioppo, 1981, 1986). The principal distinction here lies in the degree of motivation and ability for change that already exists within the target group.

When individuals have high levels of self-motivation and an intrinsic ability to engage with the message, the model suggests that a central processing route occurs, where attitude change is brought about as a result of 'mindful attention' paid to the content of a persuasive message, elaboration of its implications and integration into an individual's own set of attitudes (Jackson, 2005). By contrast, 'peripheral processing' occurs when the individual's motivation and/or ability to engage with the issue is low. In order to increase the attractiveness of adopting a change in attitude and behaviour ('source attractiveness') peripheral 'persuasion cues' can be used to functional effect. These include celebrity endorsement (for example, of a particular pro-environmental behaviour, such as opting for an energy efficient appliance), where the main motivation to engage stems from the peripheral suggestion that there are potential rewards associated with the target behaviour (often quite separate from the intended purpose of the target behaviour per se); in this example aligning someone's behaviour to that of a famous person who they respect or desire to be like (Jackson, 2005).

It is suggested that while central processing is most likely to result in long term attitude change, there are ways in which peripheral processing can also bring about enduring attitude and behaviour change. Sometimes peripheral processing can lead directly to behaviour change with altered attitudes following later.[1] The complementary elements of trust and knowledge are critical in the diffusion of information and social signals in promotion of modified patterns of behaviour; a point that has been repeatedly validated by programmes of community energy conservation, where the dissemination of information through existing social networks is a basic principle upon which their success relies (Darley and Beniger, 1981; Stern et al., 1986; Gardner and Stern, 1996).

Social Learning Theory

Social learning theory (Bandura, 1973, 1977) emphasizes the ways in which behaviour is influenced through the observation of a variety of social models, including the behaviour of parents, friends and those portrayed in the media. In addition to modelling behaviour on (what a person perceives to be) the desirable behaviour of others, the behavioural responses of social models are also particularly important elements of social learning. Positive signals (for example, observing the pleasure that someone experiences from certain behaviours) are likely to have a persuasive impact on the observer's behavioural choice. Additionally it is suggested that observing the behaviours of 'anti-role models' (that is, the negative consequences from other people's behaviour and those with whom we do not want to be associated) represents a process of learning how not to behave (Jackson, 2005).

Behavioural modelling is particularly important to the establishment and maintenance of social norms, and thus has clear implications for the effective deployment of community-based action programmes, due to the implication that one person's behaviour can have such a profound influencing effect on another's actions. Connecting with a wide range of community members in the adoption of a pro-environmental behaviour (for example, participation in kerb-side recycling or improved home insulation) is likely to increase the number of 'role models' that are able to resonate with a range of people and lifestyles and, in theory, enhance the desirability of adoption among larger numbers of community members. The role of government leadership in advancing social behavioural change is another core theme towards which social learning theory draws particular attention (that is, including a need to be seen to 'practice what they preach' and to exemplify the possibilities enabled by adoption of the behavioural change being promoted (Jackson, 2005)).

Social Norms

The concept of 'social norms' has been developed through the discipline of social psychology in an attempt to inform our understanding of the social nature of human behaviour. Cialdini et al. (1991) distinguish between the continual influences of 'descriptive' and 'injunctive' social norms. Descriptive social norms provide us with information about what people around us normally do; that is, enable individuals to fit in with regular patterns of observed behaviour. By contrast injunctive social norms imbue the individual with a sense of how others around them think that they ought to behave, reflecting the moral rules and guidelines of the social group.

Institutional rules play a key role in constraining the behaviour of individuals, largely because of the way in which social norms are often embedded in institutions (Jackson, 2004). For instance, the decision to adopt certain pro-environmental behaviours is likely to depend as much upon the existence of appropriate infrastructure and local facilities for engaging in this action as it is on positive attitudes. The availability or unavailability of reliable public transport, for example, inevitably places constraints on travel choices. It is also important to recognize the inherent limitation of choice that social norms and institutional constraints can place on individuals as consumers, as Jackson (2004) notes:

> some of these social institutional arrangements are the result of long-term cultural trends and deeply embedded social expectations ... consumers are a long way from being willing actors in the consumption process, capable of exercising either rational or irrational choice in the satisfaction of their own needs and desires. More often they find themselves 'locked in' to unsustainable patterns of consumption, either by social norms which lie beyond individual control, or else by the constraints of the institutional context within which individual choice is negotiated. (p. 1039; see also Sanne, 2002).

Personal norms refer to the things people feel obliged to do without considering what others are doing, or what they might expect. These, together with personality and situation, all have a part to play in determining an individual's response to the broader, extant social norms. Community-based initiatives that aim to engage individuals in the adoption of more sustainable, low carbon lifestyle choices therefore operate largely within the context of descriptive and injunctive social norms; their presence and influence inevitably impacting on the degree to which this engagement (and participation) is likely to prove successful.

COMMUNITY ACTION ON CARBON REDUCTION: THE UK POLICY CONTEXT

The UK Government's *Community Action 2020 – Together We Can* was announced in the 2005 Sustainable Development Strategy and has the stated aim of re-energizing action in communities across England 'to achieve a step change in the delivery of sustainable development ... by promoting new and existing opportunities to enable, encourage, engage and exemplify community action to increase sustainability' (HM Government, 2005, p. 29). This approach highlights community engagement in governance as a central facet of a sustainable society (Seyfang and Smith, 2006) and pinpoints several areas where learning and behavioural change are considered most likely to be effective through the agency of community groups. These include: tackling climate change; development of energy and transport projects; waste minimization; improvement of the quality of the local environment; and the promotion of fair trade and sustainable consumption and production.

It has been argued that *Community Action 2020* – in the context of the broader Sustainable Development Strategy – represents an increasing policy focus on the social economy as a source of sustainability transformation, active citizenship and public service delivery that incorporates social enterprise as well as community and voluntary organizations (Seyfang and Smith, 2006). Seyfang and Smith state that this was the first attempt in UK policy to address the significance of *social structures* and the need to understand the cultural and social influences that shape consumption choices, habits and impacts. This significance from a policy perspective has been brought right up to date with the publication in July 2009 of the *UK Low Carbon Transition Plan*, setting out the UK's national strategy for climate and energy (DECC, 2009). Community action is an integral component of the strategy, clearly signposted in the proposition that: 'We often achieve more acting together than as individuals. The role of the Government should be to create an environment where the innovation and ideas of communities can flourish, and people feel supported in making informed choices, so that living greener lives becomes easy and the norm' (DECC, 2009, p. 92).

The strategy also emphasizes the role for local government in this drive towards effective community engagement, clearly articulating the Government's desire to 'unlock greater action by local authorities in identifying the best potential for low carbon community-scale solutions in their areas' (DECC, 2009, p. 79). Further, it is stated that community members should increasingly be able to rely on their local authority 'to co-ordinate, tailor and drive the development of a low carbon economy in their area' (p. 94). This potentially pivotal position for local government builds on similar messages embedded in a range of other recent legislative documents and policy communications. These include the 2007 Energy White Paper, the 2006 Climate Change Programme and the new Local Performance Framework.

TRANSLATING CONCEPTS OF SOCIAL ORGANIZATION INTO PRACTICE: OPPORTUNITIES AND CHALLENGES FOR LOCAL GOVERNMENT

Drawing on recent research carried out with three English local authorities, this section considers the concepts of community organization discussed earlier in the context of some recently established engagement initiatives, looking at opportunities, barriers and challenges that have emerged in the light of their operation. The three local authorities in question (Shropshire County Council, the London Borough of Richmond-upon-Thames Council and the London Borough of Islington Council) are all currently developing a number of projects that aim to engage their communities on a path to more sustainable environmental and energy-related futures under their emerging Local Strategic Partnerships and climate change strategies.

The empirical work involved in-depth interviews with two key officers from each local authority in order to gain a 'project management' perspective on the rationale and operation of the initiatives. A brief description of the three initiatives considered in this section is provided in Table 1.1. ·

It is important to remember that for these local authorities, the motivation to address the complexities of climate change is not purely reliant upon the drive provided by central government. This position juxtaposes the broader lack of robust evidence to demonstrate the impact on carbon emissions of local or regional action, as described in CSE (2005) and goes some way to explaining why local authorities like these represent the 'exception' rather than the 'norm'. In its UK-wide review of local authority action on climate change CSE (2005) demonstrates clearly that the few current examples of good practice 'are principally down to the work of enthused, informed and committed individuals . . . [applying] their willpower, doggedness and professional expertise to create conditions within their organisation in which they can operate effectively' (p. 20). Even for 'committed' local authorities the challenge of reaching and influencing broader sections of the community via the types of engagement programme outlined in Table 1.1 remains problematic for a variety of reasons which are explored in more detail later in this section.

Opportunities

Disseminating educational messages on the growing urgency of climate change across the community, with the intention of instigating behaviour change, was agreed upon by all interviewees to be a central aim of their initiatives. For example, as these interviewees pointed out:

> Getting our community members to understand that their individual actions can have a significant impact when taken together with the actions of their neighbours and the broader community is certainly a key priority embedded in the philosophy behind our climate change projects. (Shropshire interviewee)

Table 1.1   *Outline details of community engagement initiatives established by three English local authorities*

Name of initiative	Operator	Central aims	Operation
Low Carbon Community Project	Shropshire County Council	To achieve significant reductions of CO_2 emissions within three local communities, involving household residents and businesses	Home energy checks, business and building audits, energy efficiency grants and 'Climate Change Months' awareness-raising activities (which include climate change pub quizzes, film shows, cartoon competitions and interactive workshops)
Green Living Centre	London Borough of Islington Council	A community resource to help people in the borough reduce their carbon emissions in and around the home	Face-to-face advice is available for visitors around four main areas: recycling, energy efficiency, biodiversity and green travel. A programme of 'one-off' events (for example, 'plastic bag amnesties') are organized to complement the Centre's drive to connect with the public, boost its profile and engender greater interest and increased visitor numbers
Emissions-based parking permit scheme	London Borough of Richmond Council	To reduce vehicle-related CO_2 emissions in the area, encourage use of other modes of transport and cars with smaller engines, and increase community awareness of the need to reduce transport emissions	The price of permits for each controlled parking zone is based on the previously existing charges together with the cylinder capacity of the vehicle and its CO_2 emissions. Second and subsequent permits for a household are charged at 50 per cent more than the first

There is massive potential for awareness raising, education and engaging people through that. (Richmond interviewee)

Parallel to this aim is an emphasis on co-opting the cohesion and drive of already established social networks and community groups. This was an opportunity highlighted as important by two of the interviewees, where one of them suggested that 'there is massive scope for propagating the message through word of mouth . . . and tapping into existing social networks and groups – like the Women's Institute, Parish Council and the Young People's Forum' (Shropshire interviewee). In this sense the initiatives are clearly in principle (and particularly from the 'project management' perspective), promoting and attempting to capitalize upon the processes of social learning.

Regarding the importance of awareness raising, there was clearly an understanding – accompanying desire amongst the interviewees – to put the message across in a way which would resonate effectively with the differing needs and priorities of their community members. As this interviewee argued, for example: 'You can engage people on climate change – you just have to do it in a particular way and over something that is relevant to individuals' day-to-day lives . . . the style and type of communication is key' (Islington interviewee). This issue is central to the broader objective of engaging directly with individuals, as pointed out by CSE (2009, p. 36), 'to stimulate significant reductions in emissions and normalise pro-environmental behaviours'.

In relation to this, there was strong agreement among the interviewees regarding the opportunities which could be exploited through modification of existing services in the promotion of attitudinal and behaviour change. Considering the progress of their emissions-based charging for parking permits scheme, the Richmond interviewee, for example, pointed out that 'I think one of the biggest achievements of the policy so far has been in raising awareness of the contribution that the individual can have through their choice of vehicle.' He argued that a predominantly economic-based scheme would also have the potential to influence attitudes in a pro-environmental way – to some extent echoing points made earlier in this chapter about the interplay (and interconnectedness) of institutional arrangements and social norms. This was re-iterated by the Islington interviewees, who suggested that combining the modification of existing planning services with new carbon reduction-focused advice has been an approach pursued keenly by their council.

The prospect of capitalizing directly upon latent concerns about climate change that already exist among community members was an issue highlighted by all interviewees. Each of these interviewees spoke of both environmental attitude surveys that they had recently conducted and also of levels of awareness on environmental issues that they had picked up through more informal conversations with residents. This interviewee pointed out for instance: 'Our preliminary research indicates knowledge of climate change among residents but a lack of knowledge on how to act. Our climate change projects constitute an opportunity to enable a practical translation of this knowledge through to action' (Islington interviewee).

This arguably points to another area of opportunity that links closely to social learning theory and persuasion. It has been capitalized upon particularly by the Green Living Centre in Islington, where a sizeable percentage of visitors are apparently interested in the design of the centre because they admire the 'look' of it. The relevance of this is that much of the initial ethos in establishing the centre was to make it as appealing as possible to a wide range of socio-demographic sectors (Hales, 2007). It was based on this line of reasoning that the prime high street location was chosen. As the centre's Principal Energy Adviser explained:

> there are several features that mean the centre integrates well with the Islington coffee shop culture; it's on Upper Street in Islington which is full of posh shops and cafes; the designers were very keen to make sure that it fits with the kind of shops and services in the vicinity. Some people are interested that we have a display cabinet with building materials and want to know where suppliers can be located.

Other features within the design of the centre that have continued to attract interest from visitors include a table fashioned from reclaimed wood and wooden floor boards that were originally reclaimed from a school in Sussex. It could be argued therefore that the importance of leading by example (discussed earlier as a key element in the context of social learning) is being put into practice by the local authority striving to ensure that nearly all the materials used in the design and fitting of the centre are sustainably sourced.

Barriers and Challenges

The ability of local authorities to promote the effective dissemination of their informational campaigns to encourage attitudinal and behavioural change can however be thwarted by apathy and indifference towards climate change among community members. This was one of the central conceptual barriers highlighted by all interviewees to a greater or lesser extent. As this interviewee pointed out: 'There may be a viewpoint, I sense, that there's little point in one person – or even a community over here – making changes to combat climate change when for example they see little evidence for commitment from other large polluting nations' (Richmond interviewee). On one level, this would appear to contradict the opportunities referred to by interviewees regarding the potential in latent awareness of climate change issues as revealed by their pre-project surveys and during more informal information-gathering activities. Alternatively, it may reflect the reality that latent awareness and/or good intentions do not automatically result in behavioural and lifestyle change, however well the possibilities for that action are assisted.

The experiences of participatory climate change projects in both Shropshire and Islington illustrate the difficulties associated with engaging larger numbers of community members. From the project management perspective it was suggested that this might be attributed in part to the hectic nature of modern ways of living where individuals often give higher priority to issues other than climate change. They can often display inertia in the sense of *wanting* to make behavioural changes. As these interviewees pointed out, for example:

> People's perceptions of time and how busy they are and how much they are preoccupied with different issues is a massive barrier. (Islington interviewee)

> We came up against a massive block: basically we weren't able to get bums on seats. And the feedback we got was 'oh no, not another climate change event', kind of thing. People just seemed to be thoroughly uninterested in the area for whatever reason. (Shropshire interviewee)

A central issue was felt to be the development of trust with community members as a key influencing factor in relation to establishing and maintaining engagement. For example, one of the interviewees argued that a project's success is largely dependent upon participation, which in turn depends on how effectively the initiative is promoted and also the extent to which the target community believes and trusts in the organization developing the project alongside its proposed benefits.

A potential barrier to developing a sufficient level of trust was thought to be related to the perception and 'image' of local authorities and their role in the community. One interviewee argued that 'it can be quite difficult for local authorities. The local authority has an image. Most people wouldn't dream of communicating with the local authority unless they had to' (Richmond interviewee). Another explained, 'our residents, bless them, can be quite critical of the council' (Islington interviewee). This confirms a view that the historical relationship between residents and their local authorities has often been characterized as one where limited trust and minimal confidence have been prevalent (Byrne, 2000). Whether provision of incentives for participation (for example, financial support towards insulation costs) constitutes a positive aspect of building a trust relationship with community members is not entirely clear. It does nevertheless provide a reason to participate additional to the anticipated environmental benefits that are central to this type of engagement programme.

Connecting effectively with the multiplicity of needs and priorities that exist in a community is a core challenge in this regard. As these interviewees observed:

> The right type of publicity and programme of events to get people interested and want to participate is very important. The old line about 'you can lead a horse to water . . .' is certainly relevant. (Shropshire interviewee)

> Local authorities have got to be more creative and innovative when it comes to climate change; think of things their communities would find interesting and talk to them about those things. (Richmond interviewee)

Inevitably, this requires some recognition of difference and diversity among individuals of the same community. Implicit in social learning theory, for instance, is the idea that there are benefits to be realized from approaches that embrace (rather than shy away from) inter-community diversity.

CONCLUDING REMARKS

Although defining the term 'community' is a persistent challenge, this chapter argues that having an appreciation of the inter-related concepts of social capital, social learning and social norms can be useful in providing a framework from which to develop carefully planned engagement strategies. This is in relation to 'community' itself and also the processes by which personal and social norms develop and the ways in which 'information' is individually and collectively processed over time. It is equally important to be clear about the social configurations that an engagement programme seeks to influence. The techniques and overall approach adopted will need to be designed accordingly in order to attract the interest and involvement of the community members with whom it aims to connect.

The potential agency and role for local government in mobilizing community action on climate change at the local level has been recognized and re-iterated in a range of UK policy developments, communications and guidance from central government. The UK Climate Change Programme, Low Carbon Transition Plan and Local Performance Framework all promote this role as an interface between citizens, local policy-making and delivery. Local government, it is argued, holds the potential to reach, influence and galvanize community activity. As local authorities begin (and in some cases, continue) to establish and operationalize a range of community-oriented initiatives in response to this 'call to arms' by central government, it is clear that this is not always a straightforward process.

The community initiatives considered in this chapter are attempts by three local authorities to bolster social capital, promote social learning and trigger the evolution of new social norms through the encouragement of shifts to lower carbon lifestyles. However, despite their efforts, these types of initiative face a series of barriers as well as opportunities in respect of community engagement. These barriers may well be symptomatic of broader problems associated with a reliance on local authorities as change agents in addressing the more intractable challenges of sustainable development. Even for local authorities committed to community engagement as a primary focus for their climate change agendas, problems of poor image and perception in the community, associated historically with low levels of trust and a lack of confidence in local government policies, stand to threaten the progress and success of these locally focused initiatives.

Nevertheless, community-oriented action certainly has the potential to be a potent force for social change. What seems to be of crucial importance is the nature of *how* that prospective change is encouraged and facilitated. Approaches to engaging communities need practical and pragmatic elements to be built in from the outset; information-intensive interventions alone are unlikely to be effective, as experience has demonstrated. Coordination of individual and collective behaviour change towards more sustainable patterns of living inevitably requires a policy mix that incorporates appropriate incentives and disincentives. Trust and knowledge are critical in the diffusion of social signals in promotion of changed behaviour patterns. Disseminating such signals through existing social networks has proven expeditious in the past; a key contributory factor in the success of community-based energy conservation projects, for example.

Communities are not homogeneous masses, but rather intricate and widely differing webs with varying degrees of interactivity, shared norms and communication. Appeals for making lifestyle changes to reduce carbon emissions must appear sufficiently relevant, worthwhile and feasible if they are to motivate community members to join in. They need to tap into concerns about climate change that already exist and provide additional motivation, raising awareness of the contribution that a combined effort can make in realizing reduced energy and carbon consumption.

ACKNOWLEDGEMENT

Some of the material presented in this chapter is adapted from S. Fudge and M. Peters (2009), 'Motivating carbon reduction in the UK: the role of local government as an agent of social change', *Journal of Integrative Environmental Sciences*, **6** (2), 103–20, with kind permission from the publishers Taylor and Francis Group, www.informaworld.com.

NOTE

1. Jackson (2005, p. 97) gives the example of an individual deciding to use public transport following the use of celebrity endorsement as a peripheral cue. The individual makes this change in behaviour without having deliberated over the choice – the cue and source attractiveness provide sufficient incentive. Having changed their behaviour the individual then starts to consider the benefits of public transport, thus initiating a process of follow-up attitudinal change. Jackson states that this sits well with Bem's (1972) perception theory which suggests that we sometimes infer what our attitudes are by observing our own behaviour.

REFERENCES

Bandura, A. (1973), *Aggression: A Social Learning Analysis*, Englewood Cliffs, NJ: Prentice Hall.
Bandura, A. (1977), *Social Learning Theory*, Englewood Cliffs, NJ: Prentice Hall.
Bem, D (1972), 'Self-perception theory', in L. Berkowitz (ed.), *Advances in Experimental Social Psychology 6*, London: Academic Press, pp. 1–62.
BERR (Department for Business, Enterprise and Regulatory Reform) (2007), *Energy Measures Report: Addressing Climate Change and Fuel Poverty – Energy Measures Information for Local Government*, London: BERR, accessed at www.berr.gov.uk/files/file41260.pdf.
Byrne, T. (2000), *Local Government in Britain – Everyone's Guide to How it All Works*, London: Penguin.
Chaskin, R.J., P. Brown, S. Venkatesh and A. Vidal (2001), *Building Community Capacity*, New York: Aldine De Gruyter.
Cialdini, R., C. Kallgren and R. Reno (1991), 'A focus theory of normative conduct: a theoretical refinement and re-evaluation of the role of norms in human behaviour', *Advances in Experimental Social Psychology*, **24**, 201–34.
CLG (Department for Communities and Local Government) (2006), *Strong and Prosperous Communities – A Local Government White Paper*, Norwich: HMSO, accessed at www.communities.gov.uk/publications/localgovernment/strongprosperous.
CLG (2007a), *Building Cohesive Communities: The Crucial Role of the New Local Performance Framework*, Wetherby: Communities and Local Government Publications, accessed at www.communities.gov.uk/documents/localgovernment/pdf/621282.

CLG (2007b), *An Overview of New Local Area Agreements*, Wetherby: Communities and Local Government Publications, accessed at www.communities.gov.uk/localgovernment/performanceframeworkpartnerships/localareaagreements/newlocalarea.

Cohen, A.P. (1985), *The Symbolic Construction of Community*, London: Tavistock.

Coleman, J.C. (1988), 'Social capital in the creation of human capital', *American Journal of Sociology*, **94** (January supplement), 95–120.

Cooper, H., S. Arber, L. Fee and J. Ginn (1999), *The Influence of Social Support and Social Capital on Health*, London: Health Education Authority.

Cote, S. and T. Healy (2001), *The Well-being of Nations: The Role of Human and Social Capital*, Paris: Organisation for Economic Co-operation and Development.

Crow, G. and G. Allan (1994), *Community Life. An Introduction to Local Social Relations*, Hemel Hempstead: Harvester Wheatsheaf.

CSE (Centre for Sustainable Energy) (2005), *Local and Regional Action to Cut Carbon Emissions: Report to DEFRA for the UK Climate Change Programme Review*, Bristol: CSE.

CSE (2007), *Mobilising Individual Behavioural Change through Community Initiatives: Lessons for Climate Change*, report by the Centre for Sustainable Energy and Community Development Exchange, London: HMSO, accessed at www.defra.gov.uk/Environment/climatechange/uk/individual/pdf/study1-0207.pdf.

CSE (2009), *Best Practice Review of Community Action on Climate Change*, Bristol: CSE.

Darley, J.M. and J.R. Beniger (1981), 'Diffusion of energy-conserving innovations', *Journal of Social Issues*, **37** (2), 150–71.

DECC (Department for Energy and Climate Change) (2009), *The UK Low Carbon Transition Plan: National Strategy for Climate and Energy*, London: DECC, and Norwich: The Stationery Office, accessed at www.decc.gov.uk/en/content/cms/publications/lc_trans_plan/lc_trans_plan.aspx.

Delanty, G. (2003), *Community*, London: Routledge.

DTI (Department of Trade and Industry) (2007), *Meeting the Energy Challenge: a White Paper on energy*, Norwich: The Stationery Office, accessed at www.berr.gov.uk/files/file39387.pdf.

Frazer, E. (1999), *The Problem of Communitarian Politics: Unity and Conflict*, Oxford: Oxford University Press.

Fudge, S. and M. Peters (2009), 'Motivating carbon reduction in the UK: the role of local government as an agent of social change', *Journal of Integrative Environmental Sciences*, **6** (2), 103–20.

Gardner, G. and P. Stern (1996), *Environmental Problems and Human Behaviour*, Boston, MA: Pearson Custom Publishing.

Greenwald, A. (1968), 'Cognitive learning, cognitive responses to persuasion and attitude change', in A. Greenwald, T. Brock and T. Ostrom (eds), *Psychological Foundations of Attitudes*, New York: Academic Press, pp. 147–70.

Hales, L. (2007), personal communication, interview with Laura Hales, Climate Change Partnership Officer and Lucy Padfield, Energy Manager, London Borough of Islington Council, September.

Halpern, D. (1999), *Social Capital: The New Golden Goose*, Cambridge University, Cambridge: Cambridge University Press.

Halpern, D. (2001), 'Moral values, social trust and inequality: can values explain crime?', *British Journal of Criminology*, **41**, 236–51.

Halpern, D., S. Aldridge and S. Fitzpatrick (2002), *Social Capital: A Discussion Paper*, London: Performance and Innovation Unit.

Healy, T. (2001), *Health Promotion and Social Capital*, conference paper presented at International Evidence for the Impact of Social Capital on Well Being, held at the National University of Ireland, Galway, Ireland.

HM Government (2005), *Securing the Future: The UK Government Sustainable Development Strategy*, Norwich: HMSO, accessed at www.sustainable-development.gov.uk/publications/pdf/strategy/SecFut_complete.pdf.

House of Commons (2009), *Programmes to Reduce Household Energy Consumption*, House of Commons Public Accounts Committee, Fifth Report of Session, 2008–2009, London: The Stationery Office, accessed at www.publications.parliament.uk/pa/cm200809/cmselect/cmpubacc/228/9780215526618.pdf.

Hovland, C. (1957), *The Order of Presentation in Persuasion*, New Haven, CT: Yale University Press.

Hovland, C., I. Janis and H. Kelley (1953), *Communication and Persuasion: Psychological Studies of Opinion Change*, New Haven, CT: Yale University Press.

Jackson, T. (2004), 'Negotiating sustainable consumption: a review of the consumption debate and its policy implications', *Energy and the Environment*, **15** (6), 1027–51.

Jackson, T. (2005), 'Motivating sustainable consumption: a review of the evidence on consumer behaviour and behavioural change', a report to the Sustainable Development Research Network, Policy Studies Institute, London.

Jackson, T. and L. Michaelis (2003), *Policies For Sustainable Consumption*, London: Sustainable Development Commission.

Jordan, A. (2006), *The Environmental Case for Europe: Britain's European Environmental Policy*, CSERGE working paper EDM 2006–11, University of East Anglia, Norwich.

Kawachi, I., B. Kennedy, K. Lochner and D. Prothrow-Stith (1997), 'Social capital, income inequality, and mortality', *American Journal of Public Health*, **87** (9), 1491–8.

Kirwan, K. (2008), personal communication, interview with Kerry Kirwan, Principal Energy Adviser, Green Living Centre, London Borough of Islington Council, May.

Levett, R., I. Christie, M. Jacobs and R. Therivel (2003), *A Better Choice Of Choice: Quality Of Life, Consumption and Economic Growth*, London: Fabian Society.

Long, T. (1998), 'The environmental lobby', in P. Ward and S. Ward (eds), *British Environmental Policy and Europe*, London: Routledge, pp. 105–18.

Mancini, J.A., G.L. Bowen and J.A. Martin (2005), 'Community social organization: a conceptual linchpin in examining families in the context of communities', *Family Relations*, **54**, 570–82.

Maniates, M. (2002), 'Individualization: plant a tree, buy a bike, save the world?', in T. Princen, M. Maniates and K. Konca (eds), *Confronting Consumption*, London: MIT Press, pp. 43–66.

Newby, H. (1994), 'Foreword', in G. Crow and G. Allan, *Community Life. An Introduction to Local Social Relations*, Hemel Hempstead: Harvester Wheatsheaf, pp. xi–xii.

ONS (Office for National Statistics) (2001), *Social Capital: A Review of the Literature*, Social Analysis and Reporting Division, London: Office for National Statistics.

Pelling, M. and C. High (2005), 'Understanding adaptation: what can social capital offer assessments of adaptive capacity?', *Global Environmental Change*, **15**, 308–19.

Petty, R. (1977), *A Cognitive Response Analysis of the Temporal Persistence of Attitude Change Induced by Persuasive Communications*, doctoral thesis, Columbus, OH: Ohio State University.

Petty, R. and J. Cacioppo (1981), *Attitudes and Persuasion: Classic and Contemporary Approaches*, Dubuque, IA: William C. Brown.

Petty, R. and J. Cacioppo (1986), *Communication and Persuasion: Central and Peripheral Routes to Attitude Change*, New York: Springer-Verlag.

Petty, R., J. Priester and P. Briñol (2002), 'Mass media and attitude change: advances in the elaboration likelihood model', in J. Bryant and D. Zillmann (eds), *Media Effects: Advances in Theory and Research*, 2nd edn, Hillsdale, NJ: Erlbaum, pp. 155–99.

Putnam, R. (1993), *Making Democracy Work. Civic Traditions in Modern Italy*, Princeton, NJ: Princeton University Press.

Putnam, R. (1995), 'Bowling alone: America's declining social capital', *Journal of Democracy*, **6** (1), 65–78.

Putnam, R. (2000), *Bowling Alone. The Collapse and Revival of American Community*, New York: Simon and Schuster.

Roseland, M. (2000), 'Sustainable community development: integrating environmental, economic and social objectives', *Progress in Planning*, **54**, 73–132.

Sampson, R.J. (2002), 'Transcending tradition: new directions in community research, Chicago style', *Criminology*, **40**, 213–30.

Sanne, C. (2002), 'Willing consumers – or locked in? Policies for sustainable consumption', *Ecological Economics*, **43** (2–3), 127–40.

Seyfang, G. and A. Smith (2006), *Community Action: A Neglected Site of Innovation for Sustainable Development?*, CSERGE working paper EDM 06–10, Norwich: University of East Anglia.

Stern, P.C., E. Aronson, J.M. Darley, D.H. Hill, E. Hirst, W. Kempton and T.J. Wilbanks (1986), 'The effectiveness of incentives for residential energy conservation', *Evaluation Review*, **10**, 147–76.

Studdert, D. (2006), *Conceptualising Community: Beyond the State and Individual*, Basingstoke: Palgrave Macmillan.

Walker, G.P. and N. Cass (2007), 'Carbon reduction, "the public" and renewable energy: engaging with socio-technical configurations', *Area*, **39** (4), 458–69.

Walker, G. and P. Devine-Wright (2008), 'Community renewable energy: what should it mean?', *Energy Policy*, **36** (2), 497–500.

Wilkinson, R. (1996), *Unhealthy Societies: The Afflictions of Inequality*, London: Routledge.

Willmott, P. (1989), *Community Initiatives. Patterns and Prospects*, London: Policy Studies Institute.

Woolcock, M. (2001), 'The place of social capital in understanding social and economic outcomes', *ISUMA Canadian Journal of Policy Research*, **2** (1), 11–17.

2. Sustainable communities: neo-tribalism between modern lifestyles and social change

David Evans

INTRODUCTION

Community, as the eminent sociologist Zygmunt Bauman reminds us, is a word with a certain 'feel' (Bauman, 2003). By conjuring up a sense of warmth, belonging and comfort, community almost always feels like a good thing. This feeling rings true when it comes to thinking about and addressing the challenges of sustainable development. One need only think about the emergence and implementation of Local Agenda 21 strategies or the enduring idea of 'thinking globally, acting locally' to see this. Indeed, locality is the seemingly appropriate scale of implementation insofar as it implies a movement beyond atomization and individualism without recourse to the unmanageable complexities that would accompany global solutions to global problems. Of course, community is not the same thing as locality but it occupies a similar context and in doing so carries connotations of empowerment, participation and ownership which in turn lend support to the idea that it is 'community' that feels right when it comes to the implementation of sustainable development policies and initiatives. Moving away from the broad agendas of sustainable development, this chapter considers the significance of community in relation to the contemporary calls for persons in relatively affluent societies to adopt ways of living that are somehow more sustainable.

Of course, if individuals are expected to move towards more 'sustainable lifestyles' one would be forgiven for thinking that community has been swept aside and that the possibility of social change rests on very privatized responses to environmental problems. By contrast, this chapter traces the ways in which different discourses of community feature in relation to real world efforts to live a 'sustainable lifestyle'. To do so, the analysis draws on ethnographic fieldwork conducted with an intentional community organized around the principles of sustainable living. It is argued that community, however defined, is an emergent property of these efforts as opposed to the starting point, or an existing framework through which change can be mobilized. Furthermore, it is argued that community is an intermittent feature of these efforts and that those individuals who are attempting to live sustainably move between attachment and detachment in order to pursue the multiple agendas that a (sustainable) lifestyle demands in the modern world. Crucially, community is understood to be important to the research participants in terms of providing a sense of moral regulation in addition to the more familiar purpose of facilitating emotional attachment. These are presented as important facets of existing attempts to live more sustainably, which in turn opens up a possible space through which to promote the uptake of sustainable lifestyles on a wider scale.

SUSTAINABLE LIFESTYLES

The language of sustainable lifestyles and debates surrounding the nature, necessity and possibility of people adopting them are becoming a key focus for media, comment and environmental policy. The contemporary signifi- cance of sustainable lifestyles emerges from a longer-standing interest con- cerning the pursuit of sustainable consumption (Jackson, 2005) and the idea that unsustainable patterns of consumption in industrialized nations are a major cause of environmental degradation (UNDP, 1998; OECD, 2002). Here, modern lifestyles seem to act as a surrogate for these unsustainable patterns of consumption and so the idea that sustainable consumption is a necessary condition for sustainable development (Sanne, 2002) trans- lates into the idea that sustainable lifestyles hold the key to social and environmental change. In the academic literature, another precedent can be found in the twin concerns of pro-environmental behaviour and behav- iour change. Indeed, studies in social and environmental psychology have explored the factors seemingly influencing pro-environmental behaviours (e.g. Schultz et al., 1995; Grankvist and Biel, 2001; Bamberg and Schmidt, 2003) and the effectiveness of intervention strategies to encourage such behaviours (Abrahamse et al., 2005). Sustainable lifestyles can be taken as the assemblage of pro-environmental behaviours across a range of social practices (Spaargaren, 2003). The concept of sustainable lifestyles (and lifestyle change), therefore, can be seen to bring together questions of sus- tainable consumption and the uptake of pro-environmental behaviour(s) in a manner that recognizes the cultural complexities of consumption and the interconnectedness of social practices.

To see this more clearly, the work of Spaargaren (2003) has demonstrated that the analysis of sustainable consumption and pro-environmental behaviour can profit from the concepts of lifestyle and social practice. The logic here is that social practices refer to the ways in which persons practically enact different activities according to existing systems of provision (cultural norms, existing technologies, infrastructural constraints) such that they are carried out competently and appropriately. Lifestyles can be understood, at the most basic level, as the assemblage of social practices that represent a particular way of life and give substance to an individual's ongoing narrative self-identity and self-actualization (Giddens, 1991). To see the links between social practices and consumption, the influential work of Alan Warde (2005) suggests that consumption 'occurs within and for the sake of practices' (p. 145) – a moment in virtually every practice rather than a practice in itself, such that it is hard to separate out the analysis of consumption from that of social practices. This argument becomes particularly pertinent when exploring sustainable consumption because virtually any social practice will involve the material consumption of natural resources (even if it does not involve the economic consumption of goods and services) and some sort of environmental impact.

While there is a good deal of work being done on explicating the concept of sustainable lifestyles (Hobson, 2002; Shove and Warde, 2002; Spaargaren, 2003; Connolly and Prothero, 2008; Evans and Abrahamse, 2009), it is worth taking a step back to consider some potential problems with using 'lifestyles' as a way to consider questions of environmental sustainability. Lifestyles are most commonly theorized in terms of a *life project*. The argument here is that in modern times, lifestyle is no longer ascribed by one's belonging to a traditional 'status group' (Bocock, 1992) such as a nation-state, occupational class or family structure. As a result of continued 'de-traditionalization', persons are increasingly free from conventions such that lifestyle becomes a project of self and a reflexively organized endeavour (Giddens, 1991) through which social practices and bundles thereof are chosen in the context of a multitude of possibilities. Crucially, it is almost impossible to separate accounts of this process from accounts of consumer culture on the grounds that playful consumerism is seen as the key medium through which persons are able to do so. As Featherstone has observed: 'Rather than unreflexively adopting a lifestyle, through tradition or habit, the new heroes of consumer culture make lifestyle a life project and display their individuality and sense of style in the particularity of goods, clothes, experiences . . . they design together into a lifestyle' (Featherstone, 1991, p. 86).

Of course, this is not the only way in which sociologists have conceptualized 'lifestyles'. For example, lifestyles have long been understood in relation to Bourdieu's notion of *habitus* in which structural dispositions function as the generative basis of unified social practices (Bourdieu, 1977, p. 72) and consumption is understood as a manifestation of these structural dispositions as opposed to a resource for fashioning an identity. Nevertheless, it is the most enduring framing of lifestyles and would certainly seem to be the one which resonates most clearly with the popular and policy imagination. Here, the term 'lifestyle' carries certain unavoidable connotations of 'modern consumer lifestyles' which serve to position it at the root of unsustainable patterns of consumption, which in turn renders the idea of 'sustainable lifestyles', as something of a contradiction in terms. More substantively, and pertinent to this discussion of community, Bauman's take (2002) on notions of lifestyle as life project suggests that in response to global problems we do not seek global solutions, nor do we even seek collective solutions; we seek biographical and individual solutions such as identity formation. Crucially, this implies that 'community' has no place in the pursuit of sustainable lifestyles; a suggestion which runs counter to exponents of sustainable communities and the tacit assumption that sustainable communities may well enable and support ways of living that are more sustainable. The remainder of this chapter is given to a discussion of the relationship between (sustainable) lifestyles and community as experienced through an in-depth empirical study of a fieldwork site that characterizes itself as an intentional community premised on sustainable living.

BACKGROUND AND SETTING

The fieldwork site in question is the pseudonymous Beechwood Court (BWC), an intentional community practising an alternative way of life premised on ecological and communal living. BWC is a registered educational charity set up for the provision of 'holistic education, managed by a resident community who live and work there'. To this end, they run retreats for families and individuals, community experience weeks, volunteer programmes and short courses. At the time of the study, the resident community was composed of 14 people – eight male and six female – who were predominantly white, middle class and aged between 25 and 50 (apart from one elderly gentleman and two children). It is an 11-acre site situated on the Devon/Dorset border about five miles away from the nearest town and at least two miles away from the nearest amenities. Without exception, every resident gave me an account of what a great location BWC is, both in terms of the site itself and in terms of where it is. Crucially, the location was celebrated by the residents by virtue of its distance – symbolic and literal – from 'mainstream' living such that BWC was presented as an attempt to do things differently and, most importantly, to live more sustainably (Evans, 2006). To this end, the residents of BWC engaged in (or at least presented themselves as engaging in) a range of sustainable practices ranging from extensive recycling and adoption of a vegetarian diet, to the generation of their own renewable energy sources, the adoption of permaculture methods and the installation of a reed bed sewage treatment system.

The methodology adopted in this study of BWC was an approach known to social scientists as ethnography. Basically, the goal of ethnography is to provide an account of a distinct culture or 'way of life' that is different from one's own. The idea is that ethnographers participate as fully as possible in the lives of those whom they study for a sustained period of time, observing what goes on, conducting formal and informal interviews, listening to what people say to one another and adopting a whole set of methods to help shed light on the focus of the research (Hammersley and Atkinson, 1983). Through this continued immersion, coupled with a degree of analytic 'distance', it is hoped that the ethnographer can make sense of behaviours, customs and values that at first glance might seem strange. Like all qualitative methods, the focus is on understanding rather than prediction or explanation, with a preference for depth over breadth. Crucially, ethnographic research seeks to understand through empathy: uncovering the context and premises on which phenomena are made meaningful instead of making sense of what is observed through recourse to one's own academic frameworks and cultural preconceptions.

For my part, I participated in life at BWC for an extended period of time as a volunteer member of the resident community which, essentially, involved working in their community (as all residents do – see below). During this time, I requested to do a variety of jobs so as to gain as broad an experience as possible, with access to as many persons as possible. I worked in the office and the kitchen as well as doing cleaning/odd jobs, roofing, wall building and gardening. In addition to this, I made myself present for all meals (including breakfast, which is a two hour self-service slot starting at 6 a.m.) and social events/workshops as well as trying to socialize with other members of the community. In effect, I was immersed in the full range of activity present within the site. Of course, it would not have been sufficient to 'hang out' as a member of the community because ethnography requires observation as well as participation. As such, I utilized systematic procedures such as note making, observations and asking questions as well as maintaining a certain personal detachment in order to make sense of the setting itself.

BEECHWOOD COURT: THE POSSIBILITIES OF COMMUNITY

As an exercise in communal living, it is reasonable to think of BWC's residents as organizing themselves into some kind of community. First impressions certainly supported this assumption. For example, every room is labelled with a friendly notice that carries a message of inclusion: '*our* bathroom', '*our* office', '*our* kitchen' and so on, just as the brochure in which they publicize their educational programmes promises an experience as a fully fledged member of '*our* community'. On another, deeper, level the residents of BWC do operate – at least to some extent – as a community. They live and work together within the spatial boundaries of the site. Every member of the resident community lives there full time due to the necessity of working there. Indeed, there is a lot of work to be done in terms of maintaining the community and ensuring it functions properly as an educational centre. Similarly, every member who lives there full time must work there full time because it would be impossible to live there and 'free ride'. As one resident put it: 'You have to work if you want to be part of the community . . . we only let people here in return for money on a short term basis . . . there has never been, at least to the best of my knowledge, a resident who just paid and didn't work.' Consequently, BWC's ability to work as a community (pardon the pun) is related to how well they can divide and allocate the work that needs to be done. As another resident argued:

> When people apply to become members of the community, we ask what their skills are, what they can bring to the community, 'cause there is no point in having five people who are great at gardening but nobody who is any good at cooking . . . if the only thing that somebody could bring was money then they are not much use to us, that isn't what we are about.

It seems that to be a member of the community, to be a participant, one must play a part. Of course, Durkheim has highlighted the importance of this in his *Division of Labour* (1893 [1964]), where he showed the possibilities of belonging and solidarity which could be brought about through mutual interdependence.

Beyond the division of labour, there are many ways that BWC can be thought of as a group of people organizing themselves along the lines of a community. That is, the residents seemed to have an emotional attachment to one another. The most shining example of this happened during my stay when a member of the resident community received some bad news. As soon as the other residents heard of this, they dropped what they were doing – one resident even drove back from town on her day off (she was called on her mobile telephone) – and did what they could to help. There were offers of Chai tea, group hugs, a shoulder to cry on, a lift to the coast for some meditation and a general sense of letting the individual concerned know that she had emotional support should she need it. There were many other, albeit less dramatic, examples such as offers to trade 'shifts' on the cooking rota so that a yoga class could be attended or calming each other down with a cup of tea in response to a minor crisis. It is important to note that this is how they constructed their sense of being a community: 'This is just what we do, we are a community. We are there for each other when we need it. No matter what . . . .'

I will now go on to argue that BWC does not always operate as a community but here, BWC can be seen to be characterized by a sense of community that is based on more than the mutual reliance that the division of labour necessitates.

MEAL TIMES AND THE FAILED PROMISE OF COMMUNITY

Against this 'promise of community', there were countless instances in which BWC was characterized by a good deal of isolation and solitude that does not sit easily with their conception of community. For example, I was told on my arrival that 'it is important that we all eat our meals together' and such was the apparent importance of this that the only room I was taken to while being shown around was the kitchen/dining room This gave me the impression that meal times would be an event wherein everyone sat around a large circular table, sharing food and being together way beyond the moment that the instrumental act of eating actually finishes. Furthermore, on my first day I worked in the kitchen and learned that every day those who were working in the kitchen prepared food for all of the residents to eat together. However, the reality did not match this expectation – neither on this day or any other. In actuality, the food was placed at the side of the room and residents got up and helped themselves (in a manner akin to a school canteen) before sitting down to eat. People did not sit down together and, in fact, it was rare for everyone to be in the room together at, or around the designated meal time. Some sat at the table whilst others sat on sofas at the other side of the large room. Those who did sit at the table tended to sit as far away from each other as possible. Virtually every meal was characterized by an uncomfortable silence and minimal interaction save for a little awkward small talk, often about work. On one particular day, it was a beautiful evening and several residents went outside to eat; however they all sat in different parts of the vast grounds and at no point did anybody suggest that it would be nice for everybody to sit outside together. There was literally nothing social about meal times. It was purely instrumental: people ate as quickly as possible, in silence and left as quickly as they could. Everybody did their own washing up and on more than one occasion, I saw residents queuing up, plate and cutlery in hand, waiting their turn to get to the sink.

Similarly, throughout the working day, there was very little to suggest that residents worked together. While they all worked onsite, the size of the grounds and the diversity of tasks meant that residents spent the majority of the working day in isolation, be it in the kitchen, in the office, in the gardens, or on the roof. People only tended to work in pairs when an experienced member of the community was supervising a volunteer or a new resident. Furthermore, most residents told me that they liked it this way. This sense of detachment was heightened when, as soon as the day finished and the evening meal had been sat through, people went off in separate directions to – in the words of one resident – 'do their own thing'. For example, when talking with a resident about the process of becoming a full-time resident, she told me that:

> You come for a week and then you come for progressively longer stays in order to figure out if you want to – and they want you to – come and stay for good . . . with each visit, you soon learn to bring what you need to be by yourself . . . it is the same for any volunteer. For me it is my music, I cannot be on my own without music.

She went on to say:

> When I first came, I thought I wouldn't be allowed my CD player and thought I would be around people all the time . . . as it happens, it is easy to be lonely here, particularly if you don't have a car to get away . . . you have a lot of time on your own to fill and you need to find ways to fill it.

Indeed, the site is organized in a way that almost necessitates this isolation. Every resident has their own room and these individual rooms are littered around the edges of the site such that they are far away from each other. Similarly, residents marked their detachments from the commun(e)ity through notions of private property and ownership. For instance, when I was first shown to the kitchen I was told in one breath to help myself 'because the food is communal' and in the next to 'avoid that shelf because the food here belongs to individual residents'. Similarly, when chatting in the library, I was told not to touch any of the books in one of the cabinets because they belonged to an individual resident who gets very angry if people touch his books.

EMERGENCE AND INTERMITTENCE

In trying to understand how 'community' features in relation to the sustainable lifestyles practised at BWC, it is hopefully becoming clear that it is not appropriate to think of BWC simply in terms of either being a community or not being a community. Indeed, the empirical material suggests that BWC is characterized by a good deal of movement between practices and processes of attachment and practices and processes of detachment. Consequently, in the context of this research, a little more theoretical explanation is required to make sense of the relationship between (sustainable) lifestyles and the idea of community.

Identity and Identification

As argued earlier, lifestyles are most commonly understood as projects of identity formation and self-actualization. It would be a mistake, however, to think of lifestyles as atomized projects of self. In many ways, the concept of lifestyles can be thought of as straddling the tension between notions of identity on the one hand and community on the other: 'Identity is about more than the development of a life project . . . it is fundamentally about issues of belonging, expression, performance, identification and communication with others' (Hetherington, 1998, p. 62).

Indeed, if the residents of BWC are to be understood as pursuing a 'life-project', they certainly seemed to intimate that the 'community' to which they belonged was an important part of this process. For example: 'Yeah, I'm finding myself here . . . I guess I was a bit lost, bit out of place before I came here . . . I was always a bit different, not really myself but being here is helping me to figure out what I am about.' To make sense of this, we can consider lifestyles in relation to (post)modern forms of sociality. There are many concepts to which we could turn, such as Turner's notion of 'communitas' (1969) or Bauman's 'peg communities' (2001), but it is Maffesoli's analysis of neo-tribalism (1996) that would seem to be most pertinent to this analysis. Maffesoli suggests that people do not belong to a single group or tribe; rather they move freely between many attachments and groupings. This was certainly true of the residents of BWC insofar as they seemed to move between attachment to and detachment from the 'community' of BWC. Furthermore, it was rare for persons to limit their attachments to the community at BWC because every opportunity was taken to detach from BWC and attach to other collectives. Interestingly, it was rare for anybody to remain a resident member of the 'official' BWC community for more than two years.

Hetherington (1994), drawing on Schmalenbach's concept of the *bund*, sheds further light on these observations. A *bund* is an elective affinity grouping that sits somewhere between notions of community and those of individuality; providing freedom from the ascriptive elements of community alongside opportunities to experience the 'transcendent warmth of the collectivity'. Consequently, it is perhaps fruitful to think of community as less of a framework through which to mobilize sustainable lifestyles and more as an emergent property of and valuable resource for their pursuit. Indeed, it makes a lot of sense to think of BWC as: (i) the voluntary and yet temporary coming together of persons attempting to live in ways that are more sustainable; (ii) a useful 'peg' through which meaning is given to the practices that this entails; while (iii) offering an experience of warmth and communion through emotional attachments to others. In putting the concept of 'lifestyles' at the centre of the analysis, they can be seen to embody notions of choice and self-actualization whilst offering opportunities for collectivity and attachment in the form of neo-tribal sociality and membership to a *bund*.

Aesthetics and Ethics

It is all too easy to think about lifestyles in terms of shallow, playful performances that represent the triumph of style over substance and aesthetics over ethics. Indeed, this is implicit in Bauman's idea that lifestyles and identity preclude the possibility of a collective (let alone global) solution to global problems. Nevertheless, there is no reason for the concept of lifestyles to exclude the provision (or at least presence) of social regulation or moral governance. For example, Giddens (1991) suggests that lifestyles can be thought of as existential projects that address issues of how one should live one's life. Similarly, the theories of Maffesoli and Hetherington make clear the links between identification and ethics. Hetherington notes that those electing affinity within a lifestyle grouping are: '[m]ore likely to seek collectives of like minded others. This is especially so in the case of those who seek to create a lifestyle that is ethically committed towards others' (Hetherington, 1998, p. 94).

He goes on to suggest that: '[l]ifestyles . . . seek to make life meaningful on affectual and value-rational grounds . . . others – indeed the whole category of Other – becomes significant on emotional *and* moral grounds' (Hetherington, 1998, p. 94, my emphasis).

The crux of Hetherington's argument is that the communities that emerge from life projects are not only about finding others with whom one can identify; they are about finding others with shared values. That is to say, the source of the group's identification is some sense of shared values. For example:

The interesting thing is that people, uh before I came here, people outside thought I was a nut for wanting to be greener. I think my family thought I was some sort of hippy but being here is good as people understand why I want to do these things. And it is a lot easier to do these things when you are around people doing the same.

For Hetherington, there is no distinction between an emotional community (having some sort of social bond with others) and a moral community because 'morality' is innately tied up with notions of emotional attachment and community. Again, this idea can be traced back as far as Durkheim (1893 [1964]), who theorized sociality – or in his terms, solidarity – as morality and/or an emotional attachment to others. Similarly, his famous study of suicide (Durkheim, 1897 [1952]) suggests that moral regulation of the individual by the community is an important facet of individual well-being which connects back to the idea that lifestyles involve the asking of questions concerning the right way to live (Giddens, 1991).

Viewing the importance of community in relation to moral regulation at BWC, it is worth noting that the movements between attachment and detachment tended to mirror the movements that residents made between social practices that are understood as sustainable and those that are not. The moments in which the residents of BWC were operating as a collectivity were also those in which they practically – or at least discursively – enacted sustainable living. Conversely, movements away from the gaze of the collective resulted in respondents engaging in practices that do not necessarily sit well with their objective of living more sustainably. For example, although meal times could hardly be described as convivial they were collectively experienced insofar as they were prepared for everybody and taken together at the same time. These meals were, unsurprisingly, 100 per cent vegetarian (often vegan) and made from local, organic and home grown ingredients.

During my time at BWC, however, I gradually learned that the food practices of residents were often radically different when they were not in the company of other BWC residents. Indeed, I witnessed and was treated to tales of residents sneaking into their bedrooms with a pork pie or a takeaway, as well as incidents of people eating meat and 'whatever the hell they like' when away from the premises. Similarly, DVD players, hair straighteners, hi-fi systems, televisions and even games consoles were ubiquitous at BWC. Of course, these items were never present in shared, communal, public or visible areas where they would serve to pollute the categorical enactment of sustainable living; they were tucked away in residents' private spaces. Most notably, and in contrast to ways in which the location of BWC was celebrated as emblematic of symbolic distance from the modern world (not to mention their desires to live more sustainably),

virtually every resident owned a car in order to 'travel back' and away
from the BWC community (Evans, 2006). Of course, everybody knew that
everybody else had a car but the car park was well removed (and hidden)
from the places in which they came together to act as a community. As one
resident told me while gesturing towards the area in which the cars were
parked: 'what a joke . . . it's like —ing Detroit out there'. So, the emergence
of community seems to facilitate the practices that represent a response to
environmental crises and the key seems to be the moral governance occa-
sioned by the gaze of the collective. Moreover, community emerges from
lifestyle insofar as movement into the collective – and the moral regulation
this entails – is voluntary and part of the life projects that the residents of
BWC are pursuing. Consequently, it is quite possible to see that the possibility of community responses to environmental challenges is not extruded
by a focus on (sustainable) lifestyles. The only problem is that the intermittence and fluidity associated with neo-tribal sociality mirror intermittence
and fluidity when it comes to adopting sustainable social practices.

DISCUSSION

In tracing how community features in relation to the sustainable lifestyles
being pursued at BWC, it is tempting to theorize what was observed in
terms of failure and contradiction. However, working with the idea that
social practices are the building blocks of lifestyles, it is worth recalling
Spaargaren's work on sustainable lifestyles in which he stresses that they
may vary considerably amongst themselves with respect to the net environmental impact of the individual's lifestyle (Spaargaren, 2003, p. 689)
such that the analysis of sustainable consumption should look across
a range of social practices without necessarily demanding consistency.
Indeed, the lifestyles experienced at BWC cannot be understood as consistently 'sustainable' across social practices but there is no denying that
there is at least some effort to live and consume sustainably (of course,
the success of these efforts is beyond the toolbox of a sociologist!). It is
perhaps interesting to note that this inconsistency is related to residents
not wanting and not being able to turn their back on modern living and
the 'mainstream' society that they are attempting to subvert (see Evans,
2006). It follows that – instead of the temptation to level charges of
hypocrisy – it is fruitful to think of the residents of BWC as attempting to
balance the demands of living modern lifestyles whilst also making efforts
to reduce their environmental impact. The interesting thing here is that
some of the limitations of the data (focused as they are on persons who
are intentionally and ostensibly attempting to live sustainably) actually

lend support to some of the wider questions addressed at the start of this chapter because the tensions facing the residents of BWC can only hint at the difficulties that might be encountered in motivating sustainable lifestyles on a broader scale.

That said, there are some potentially important things to take from the analysis of BWC insofar as the occasioning of sustainable social practices is itself tied to the emergence (albeit intermittently) of community. At BWC, the idea of community is conceived of as a broadly positive phenomenon and is experienced as an important part of residents' attempts to live a sustainable lifestyle. This is true in terms of providing emotional attachments and identification with others but also in terms of providing scope for moral regulation and answering questions about the right way to live. Exploring these issues through the lens of neo-tribalism implies all of the above without demanding consistency. Those wishing to promote sustainable lifestyles – academics and policy makers for instance – should perhaps consider using the promise and possibility of community as a useful tool for doing so, although a more thorough discussion of what this might entail is beyond the remit of this chapter. Indeed, emphasizing the opportunities for belonging and offering refuge from moral ambivalence without the demand for consistency or wholesale involvement might represent a useful point of entry to motivate the uptake of lifestyles that straddle the demands of modern living and the need for social and environmental change. Viewed as such, the idea of community retains the 'feel' with which this chapter opens but it is the twin concepts of lifestyle and life project that enable – conceptually and practically – more isolated consideration of these issues.

REFERENCES

Abrahamse, W., L. Steg, C. Vlek and T. Rothengatter (2005), 'A review of intervention studies aimed at household energy conservation', *Journal of Environmental Psychology*, **25**, 273–91.

Bamberg, S. and P. Schmidt (2003), 'Incentives, morality, or habit? Predicting students' car use for university routes with the models of Ajzen, Schwartz, and Triandis', *Environment and Behavior*, **35**, 264–85.

Bauman, Z. (2001), 'On mass, individuals and peg communities', in N. Lee and R. Munro (eds), *The Consumption of Mass*, Oxford: Blackwell, pp. 102–13.

Bauman, Z. (2002), *Society Under Siege*, Cambridge: Polity Press.

Bauman, Z. (2003), *Community: Seeking Safety in an Insecure World*, Cambridge: Polity Press.

Bocock, R. (1992), 'Consumption and lifestyles', in R. Bocock and K. Thompson (eds), *Social and Cultural Forms of Modernity*, Cambridge: Polity Press, pp. 353–64.

Bourdieu, P. (1977), *Outline of a Theory of Practice*, Cambridge: Cambridge University Press.

Connolly, J. and A. Prothero (2008), 'Green consumption: life politics, risk and contradictions', *Journal of Consumer Culture*, **8** (1), 117–46.

Durkheim, E. (1893 [1964]), *The Division of Labour in Society*, New York: Free Press.

Durkheim, E. (1897 [1952]), *Suicide*, London: Routledge.

Evans, D. (2006), 'An ethnography of alterity: margins, markets, morality', PhD thesis, University of Cardiff.

Evans, D. and W. Abrahamse (2009), 'Beyond rhetoric: the possibilities of and for "sustainable lifestyles"', *Environmental Politics*, **18** (4), 486–502.

Featherstone, M. (1991), *Consumer Culture and Postmodernism*, London: Sage.

Giddens, A. (1991), *Modernity and Self Identity*, Cambridge: Polity Press.

Grankvist, G. and A. Biel (2001), 'The importance of beliefs and purchase criteria in the choice of eco-labelled food products', *Journal of Environmental Psychology*, **21**, 405–10.

Hammersley, M. and P. Atkinson (1983), *Ethnography: Principles in Practice* 2nd edn, London and New York: Routledge.

Hetherington, K. (1994), 'The contemporary significance of Schmalenbach's concept of the Bund', *Sociological Review*, **42** (1), 1–25.

Hetherington, K. (1998), *Expressions of Identity: Space, Performance, Politics*, London: Thousand Oaks, and New Delhi: Sage.

Hobson, K. (2002), 'Competing discourses of sustainable consumption: does the rationalisation of lifestyles make sense?', *Environmental Politics*, **11** (2), 95–119.

Jackson, T. (2005), *Motivating Sustainable Consumption: A Review of Models of Consumer Behaviour and Behaviour Change*, report to the Sustainable Development Research Network, London: Policy Studies Institute.

Maffesoli, M. (1996), *The Time of the Tribes: The Decline of Individualism in Mass Society*, London: Thousand Oaks, and New Delhi: Sage.

OECD (Organisation for Economic Co-operation and Development) (2002), *Towards Sustainable Household Consumption? Trends and Policies in OECD Countries*, Paris: OECD.

Sanne, C. (2002), 'Willing consumers – or locked in? Policies for a sustainable consumption', *Ecological Economics*, **42** (1–2), 273–87.

Schultz, P.W., S. Oskamp and T. Mainieri (1995), 'Who recycles and when? A review of personal and situational factors', *Journal of Environmental Psychology*, **15**, 105–21.

Shove, E. and A. Warde (2002), 'Inconspicuous consumption: the sociology of consumption, lifestyles and the environment', in R. Dunlap, F. Buttel, P. Dickens and A. Gijswijt (eds), *Sociological Theory and the Environment: Classical Foundations, Contemporary Insights*, New York and Oxford: Rowman and Littlefield, pp. 230–51.

Spaargaren, G. (2003), 'Sustainable consumption: a theoretical and environmental policy perspective', *Society and Natural Resources*, **16**, 687–701.

Turner, V. (1969), *The Ritual Process: Structure and Anti-Structure*, London: Routledge.

UNDP (United Nations Development Programme) (1998), *Human Development Report*, Oxford: Oxford University Press.

Warde, A. (2005), 'Consumption and theories of practice', *Journal of Consumer Culture*, **5** (2), 131–53.

3. The social dimensions of behaviour change: an overview of community-based interventions to encourage pro-environmental behaviours

Wokje Abrahamse

INTRODUCTION

Imagine you are walking down a parking lot and find a promotional leaflet underneath the windscreen wipers of your car. The parking lot is scattered with discarded leaflets. What are you likely to do with the leaflet? Would you walk to the nearest rubbish bin, or would you throw it away on the spot? Now imagine the parking lot is clean, and clear of leaflets. What are you likely to do in that case? Admittedly, the first situation, where leaflets are scattered around a parking lot, is sending out a message that it is 'OK' to litter, whereas the other situation is sending out quite a different message. How do people behave as a result of such social and normative influences? How can such processes be employed to promote behaviour change in a more environmentally friendly direction?

When the consequences of our individual actions are visible to other people, normative or social influences are important factors to take into consideration. In fact, a large body of research exists in social and environmental psychology, exploring the relationships between social processes on the one hand and behaviour change on the other. If the aim is to encourage people to change their behaviour in a more environmentally friendly direction, be it on an individual or group/community level, it is necessary to examine these underlying factors. This chapter aims to provide an overview of the psychological literature on the role of community-based interventions, with a specific focus on social and normative influences in understanding and motivating behavioural change. It will provide an overview of some key concepts, as well as an overview of community-based intervention studies. More specifically, the chapter will focus on the role of social norms, social identities and social comparison processes as potential motivators for change. As such, this chapter will combine findings from theory and practice, and provide recommendations for intervention planners and policy makers.

THEORETICAL APPROACHES: SOCIAL INFLUENCES AND BEHAVIOUR CHANGE

How we behave is strongly influenced by the behaviour of the people around us. By talking and interacting with other people, we form our opinions, and our beliefs about how we ought to behave, or about what is – for want of a better term – 'socially desirable', such as being silent in library settings (Aarts and Dijksterhuis, 2003). Such social or normative influences are important factors to take into consideration, especially when the outcomes of our individual behaviour are visible to others. As our behaviour takes place in (various) social contexts, our behaviour is often guided by social and normative influences and, as such, they may form interesting entry-points for interventions aimed at encouraging pro-environmental behaviour change.

Social Norms

Social norms are conceived of as sets of beliefs about what other people are doing, or what they approve or disapprove of doing. As referred to in Chapter 1, a distinction can be made between descriptive and injunctive social norms (Cialdini et al., 1991). Descriptive social norms refer to what is considered the 'normal' thing to do (that is, beliefs about what the majority of people are doing in a given situation), while injunctive social norms refer to what 'ought' to be done (that is, beliefs about what other people think we should or should not be doing in a given situation). Social norms are formed through social interaction and guide our behaviour, particularly when these norms are activated. The littered parking lot mentioned in the example at the beginning of this chapter may activate a norm in favour of littering, which in turn may inform how we behave in such settings. In a series of field studies, Cialdini and colleagues (Cialdini et al., 1991) found that littering behaviour depended on the amount of litter that was present in the environment. Leaflets with appeals to refrain from littering (that is, injunctive social norm) were distributed in environments with varying degrees of litter. Results indicated that the leaflets were discarded more often in littered environments (a descriptive norm in favour of littering: 'it is OK to litter here, other people are doing it') as opposed to non-littered environments ('it is not OK to litter').

Social Identities and Social Comparison

As individuals, we are part of many different social categories and social groups. Social identity theory (Tajfel and Turner, 1986) has been developed to understand and explain intergroup behaviour. According to social identity theory, people tend to strive for a positive self-image. There are two processes by which people can obtain a positive image of themselves. First, by means of emphasizing one's social identity, or membership of a certain group. A social identity is often defined as 'that part of the self concept that derives from one's membership of social groups' (Hogg and Vaughan, 2002), and it refers to shared meanings of who one is and how one should behave. People who strongly identify with a certain social group will also be more likely to act in accordance with the norms of that group. For instance, Terry et al. (1999) found that a stronger social norm in favour of recycling (the extent to which friends and peers were in favour of recycling) was related to stronger intentions to recycle, but only for those who (more strongly) identified with this reference group (friends and peers). For those respondents who did not identify strongly with their group of friends and peers, the relationship between social norms and intentions to recycle was weaker. This finding suggests that when social identity (or group membership) is made salient, social norms can form the basis for behavioural choices, depending on the extent to which people identify with the group, thereby providing interesting avenues for research on pro-environmental behaviour change.

The second process that can be distinguished here is social comparison. People have a general tendency to compare themselves with other people (for example, in terms of achievements) in order to make themselves feel good (although some people have stronger tendencies to do so than others). This so-called social comparison refers to the process of thinking about information about one or more other people in relation to oneself (Wood, 1996). For instance, comparing oneself with others may act as a source of inspiration and may enable people to improve their current situation. Alternatively, it may act as a stimulus to change behaviour (for example, feelings of competition). As such, comparison processes can be employed as part of interventions to promote behaviour change (for example, feedback about the performance of others).

PRACTICAL APPROACHES: COMMUNITY-BASED INTERVENTIONS

The social and environmental psychology literature reveals an abundance of intervention studies with an aim to encourage consumers to adopt more environmentally friendly behaviours – with varying degrees of success (see Dwyer et al., 1993; Schultz et al., 1995; Abrahamse et al., 2005). However, as has been noted elsewhere (Gardner and Stern, 2002), the number of (published) studies that evaluate the effectiveness of community-based interventions is relatively low.

Community-based interventions take place within a certain social context (for example, neighbourhoods) and generally focus on employing social and normative influences to motivate behavioural change. Ideally, community-based approaches first examine the potential barriers to change and then design and implement the interventions accordingly (McKenzie-Mohr, 2000). In this section, an overview is provided of community-based interventions. The following interventions will be discussed: normative information, public commitment and group/comparative feedback. It is beyond the scope of this chapter to provide an exhaustive overview. Other, more extensive, reviews of the intervention literature can be found elsewhere (for example, Dwyer et al., 1993; Schultz et al., 1995; Abrahamse et al., 2005).

Information

Information provision is widely used to encourage the adoption of environmentally friendly behaviours – its success, however, is debatable. Such information (for example, about energy conservation) generally leads to an increase in knowledge, or awareness, but it does not necessarily translate into behaviour changes (Gardner and Stern, 2002; Abrahamse et al., 2005).

It has been suggested that the provision of information may be more effective when it makes salient social norms in favour of pro-environmental behaviours. One particular study monitored towel use and re-use in a hotel setting (Goldstein et al., 2008). Towels were re-used more frequently by hotel guests when they were provided with information that emphasized descriptive social norms in favour of reuse (for example, 'did you know 75 per cent of our guests help save the environment by re-using their towels') compared with when they were given information that contained environmental norms only (for example, 'help save the environment by re-using your towels'). This suggests that activating a social norm in favour of a certain pro-environmental behaviour can be an effective way to encourage this behaviour.

Information provision can also be more effective when it is given in a certain social context. Neighbourhood interactions may be important in this respect, as these may facilitate the diffusion of information and may help people develop and establish social norms in favour of environmentally friendly behaviours. A study by Weenig and Midden (1991) evaluated the effectiveness of an information campaign aimed at encouraging the adoption of energy-saving measures (for example, insulation). This study found that awareness of this campaign was higher when people had more ties in their neighbourhood. Additionally, the decision to adopt such measures was related to the strength of these neighbourhood ties: advice received from strong ties was related to adoption decisions while advice received from weak ties was not.

Information provision has also been evaluated in organizational settings. One particular study was aimed at encouraging petrol conservation (by means of a more fuel-efficient driving style) among employees of the Dutch postal service (Siero et al., 1989). First, a pilot study was conducted that aimed to examine employees' perceptions of their driving behaviour and barriers to altering their behaviour. Then, the intervention programme was designed with the issues that had been picked up in the pilot in mind. For instance, the information addressed the apparent misperception that energy-efficient driving meant a loss of time. Compared with a control group, drivers who had been exposed to the interventions (information and feedback) reduced their petrol consumption significantly. In addition, drivers in the experimental group perceived stronger social norms in favour of conservation, highlighting the role that social norms may play in behaviour change programmes.

Another approach to enhancing the effects of information is referred to as the 'block leader' approach. This approach usually entails recruiting volunteers in a certain neighbourhood who will act as 'opinion leaders' in helping to encourage certain pro-environmental behaviours. Hopper and Nielsen (1991) evaluated the block leader approach to promote participation in a community-wide recycling programme in an urban neighbourhood. In one part of the neighbourhood, block leaders were recruited, and they were asked to encourage neighbourhood residents to recycle, by providing information about the community recycling programme and by handing out reminders just before recycling pick-up dates. Households in two other parts of the neighbourhood either only received information or only received a reminder before the pick-up dates. At the end of the seven-month intervention period, the block leader approach appeared to have been most effective in encouraging recycling. Social norms in favour of recycling were more strongly endorsed by households who had been part of the block leader programme, suggesting a positive influence of neighbourhood interactions. Taken together, the results of these studies indicate the importance of normative influences in promoting pro-environmental behaviours.

Public Commitment

Commitments are potentially powerful and cost-effective interventions. A commitment entails making a promise to try and engage in a certain pro-environmental behaviour (for example, try driving less often, start recycling, etc.). In doing so, a moral obligation may be activated to stick to the promise (that is, personal norms; Schwartz, 1977). Commitments can either be done privately, to oneself, or they can be made publicly, in a leaflet for instance. Public commitments may act as a social motivator to change behaviour (for example, avoid a negative image).

Pallak and Cummings (1976) compared the effects of private and public commitments to promote energy conservation among households. Those who had signed a public commitment (that is, publication in a leaflet) showed a lower rate of increase in energy consumption than those in either the private commitment or the control group. This effect was maintained over a period of six months following discontinuation of the intervention. This suggests that a commitment – through eliciting a disconnect between the promise and actual behaviour – can be a powerful tool for change, as its effect may last even after the initial promise was made. Various studies have looked at the effects of public commitments to encourage recycling (for example, Burn and Oskamp, 1986; Wang and Katzev, 1990; DeLeon and Fuqua, 1995). Generally speaking, the results of these studies point to the relative success of public commitments, although exceptions do exist (see Burn and Oskamp, 1986). DeLeon and Fuqua (1995) evaluated the effectiveness of a community-based initiative to promote recycling. Residents either received feedback on how much they had recycled as a group (that is, in terms of pounds of recycled paper) or they were asked to sign a public commitment to recycle (their names would be published in a local newspaper). A third group received a combination of public commitment and group feedback. This combined-intervention group was the only group to recycle significantly more compared with baseline levels (prior to intervention). However, the effects were not measured over longer periods of time, so the durability of these effects is unknown.

Research by Wang and Katzev (1990) found that the effects of group commitment on recycling were different for different target groups. One of their studies was conducted in a retirement home, and here they found that residents who had signed a public commitment increased the amount of recycled paper, both during the intervention as well as through a follow-up period – corroborating that commitments can be effective beyond the actual intervention period (cf. Pallak and Cummings, 1976). It should be noted that no control group was used in this study, and the sample was relatively small (N = 24). In contrast, a second study – conducted on campus, with 87 students – found that a public commitment was less effective than an individual commitment. The authors attribute these different effects for group commitment to potential differences in interaction and group cohesion in the two settings. The results of these studies highlight the potential for public commitments to encourage (lasting) behavioural changes. However, these studies did not include measures of personal or social norms, or other potentially relevant variables, and therefore relatively little is known about how (public) commitments work.

Group/Comparative Feedback

Feedback about individual performance relative to the performance of others may be helpful in encouraging pro-environmental behaviour change. By providing people with feedback on how they are doing as a group, social norms in favour of a certain pro-environmental behaviour may become salient. When comparisons are made with the performance of other groups, social comparison may be evoked, especially when important or relevant others are used as a reference group. The results of the use of comparative feedback are mixed, and seem to be dependent on the target group and the target behaviour that are studied.

Studies in the area of recycling seem to suggest that providing group feedback is an effective strategy. For instance, Schultz et al. (1995) evaluated the effect of group feedback in an existing community-wide recycling scheme. This group feedback consisted of feedback about the average amount of recycled materials and the average participation rates in the neighbourhood (that is, descriptive social norm). Results indicate that recycling increased among households who received group feedback (and also among households who received individual feedback). Providing households with information about recycling or a plea to recycle did not have the desired effect. This indicates that descriptive norm feedback is a potentially effective strategy to encourage recycling among householders. It should be noted however that this study did not measure whether descriptive norms in favour of recycling changed as a result of the intervention.

Comparative feedback is also part of the so-called EcoTeam Program (ETP). EcoTeams are small groups (for example, neighbours, friends, family) who come together once every month to exchange information about energy-saving options. They also receive feedback about their own energy savings and the savings of other EcoTeams. A Dutch study (Staats et al., 2004) found that the ETP approach was successful in reducing energy use and in increasing the frequency of energy-saving behaviours, both shortly after the programme and during a follow-up two years later.

Based on these results, such a community-based approach appears to be quite successful, also in the long run.

Another line of research suggests that the effects of comparative feedback may depend on whether people already behave according to the norm. For instance, Schultz et al. (2007) presented one group of households with feedback about their daily electricity consumption which contained descriptive norm information (that is, about the average electricity consumption in their neighbourhood), and a second group of households received descriptive as well as injunctive norm feedback (that is, information about whether the household had done better or worse than the neighbourhood average). For relatively high electricity consumers, descriptive norm feedback resulted in energy savings. However, for relatively low electricity consumers (that is, those who already adhered to the norm), the same descriptive norm feedback resulted in increased energy consumption. When this descriptive norm feedback was combined with comparative feedback, these low electricity consumers maintained their low levels of energy consumption. These results indicate that for those people who already behave in a pro-environment fashion, descriptive norm feedback may in fact have unintended effects.

Group/comparative feedback has also been implemented in organizational settings (for example, Siero et al., 1996; Staats et al., 2000). The study by Siero et al. (1996) used a combination of information, goal setting and comparative feedback to encourage employees to engage in energy-efficient behaviour on the job. Employees who had received comparative feedback (about how they were doing compared to another company) saved significantly more energy than employees who had not. After six months the difference between the two groups was still present, indicating the long term potential of comparative feedback in a work setting. Interestingly, these behaviour changes were not accompanied by changes in social norms in favour of conservation.

CONCLUSIONS AND RECOMMENDATIONS

Many environmental problems, such as energy use, are related to human behaviour and, consequently, may be reduced through behavioural changes. This chapter provides an overview of the social dimensions of behaviour change, focusing on community-based interventions. In this concluding section, several guidelines are proposed to help researchers and policy makers effectively design, implement and evaluate community-based intervention programmes to encourage more environmentally sustainable behaviour patterns.

Social and normative influences appear to be important factors in encouraging behaviour change. For instance, one study found that information that made social norms in support of pro-environmental behaviours salient was successful in encouraging behaviour change, compared with information that only contained 'standard' environmental information (Goldstein et al., 2008). There is also some indication that certain interventions using social or normative influences have an effect that lasts even after the intervention has been discontinued. To illustrate, interventions that take place within a social setting, such as EcoTeams, or the block leader approach, can provide lasting effects (for example, Hopper and Nielsen, 1991; Staats et al., 2004). The success of a block leader approach for instance may be attributable to the fact that these block leaders are positioned within a community, and encourage the diffusion of information or the development of social norms in favour of pro-environmental behaviour. Interestingly, evaluation studies of community-based approaches are largely underrepresented in the environmental psychology literature. Future studies could focus on employing social and normative influences as part of interventions.

It should be noted that certain interventions discussed in this chapter do appear to have differential effects according to the target group, or the target behaviour in question. For instance, comparative feedback was more effective than individual feedback in encouraging energy-efficient behaviour in the workplace (for example, Siero et al., 1996), but this is not the case for household settings, where generally no differences are found between individual and comparative feedback (for example, Midden et al., 1983; Abrahamse et al., 2007). Similarly, a study on recycling behaviour found that group commitment was effective in encouraging recycling in a retirement home but not in among university students living on campus (Wang and Katzev, 1990). Interventions such as public commitment or group feedback may be especially effective when there is regular interaction and/or a strong cohesion among group members. On a different but related note, social norm information appeared to have different effects, depending on whether people already adhered to the norm or not (see Schultz et al., 2007). This has also been found in a number of other studies, in that high consumers of energy reduced their energy use while low energy consumers increased their energy use as a result of individual feedback provision (see Abrahamse et al., 2005 for an overview). This is an important finding from a policy perspective, highlighting that interventions that make a social norm or benchmark salient may actually not have the intended effect for those who are already behaving in a pro-environmental way.

Based on the overview of the literature, several guidelines can be proposed to effectively develop, implement and evaluate community-based interventions. An important step in designing interventions aimed at encouraging behaviour change is a thorough problem diagnosis (McKenzie-Mohr, 2000; Geller, 2002). This could be done by identifying factors that make sustainable behaviour patterns (un)attractive (for example, social norms). It is important that interventions address and change possible barriers to behavioural change (see also Gardner and Stern, 2002). In terms of reducing environmental impact, it is important to identify those behaviours that have a relatively large impact. By keeping environmental goals in mind, researchers and intervention planners can focus on behaviours and target groups that significantly influence environmental qualities.

When evaluating the effects of a community-based intervention, it is important to monitor changes in behaviour as well as behavioural determinants. Many studies discussed here reveal only to what extent interventions have been successful, without necessarily providing insight into the reasons why. Did social norms change as a result of the provision of normative information? Did neighbourhood residents start comparing themselves with their neighbours as a result of the provision of comparative feedback? These insights are important for the further development of interventions. Say for instance that group feedback did not result in the anticipated energy savings; this may well be attributable to the fact that social norms did not change as a result of the intervention, or it may well be that people were not susceptible to normative influences because they did not identify very strongly with the group (for example, Terry et al., 1999). In other words, ideally the effectiveness of interventions and the determinants of behaviour should be examined simultaneously, as this can provide input for the further improvement of interventions.

The final recommendation is related to the observation that intervention studies typically have a mono-disciplinary focus. For instance, sociologists can provide valuable insight into the meanings individuals attach to sustainable practices, with regards to existing institutional and contextual arrangements (Spaargaren, 2003). Also, input from environmental scientists can be of valuable importance to further improve intervention studies. The environmental sciences can help translate energy-related behaviours into their environmental impact, for example in terms of CO_2 emissions, and help select high impact behaviours. It is therefore important to consider interventions to promote sustainable behaviour patterns from an interdisciplinary perspective. Equally well, close collaboration between academia and the policy arena is essential in order to develop and evaluate effective interventions. The guidelines proposed here may help researchers and policy makers to design and implement effective community-based approaches to encourage the adoption of environmentally friendly behaviours.

REFERENCES

Aarts, H. and A. Dijksterhuis (2003), 'The silence of the library: environment, social norm, and social behavior', *Journal of Personality and Social Psychology*, **84**, 18–28.

Abrahamse, W., L. Steg, C. Vlek and T. Rothengatter (2005), 'A review of intervention studies aimed at household energy conservation', *Journal of Environmental Psychology*, **25**, 273–91.

Abrahamse, W., L. Steg, C. Vlek and T. Rothengatter (2007), 'The effect of tailored information, goal setting, and tailored feedback on household energy use, energy-related behaviors and behavioral antecedents', *Journal of Environmental Psychology*, **27**, 265–76.

Burn, S.M. and S. Oskamp (1986), 'Increasing community recycling with persuasive communication and public commitment', *Journal of Applied Social Psychology*, **16**, 29–41.

Cialdini, R.B., C.A. Kallgren and R.R. Reno (1991), 'A focus theory of normative conduct: a theoretical refinement and reevaluation of the role of norms in human behavior', in M.P. Zanna (ed.), *Advances in Experimental Social Psychology*, vol 24, San Diego, CA: Academic Press, pp. 201–34.

DeLeon, I.G. and R.W. Fuqua (1995), 'The effects of public commitment and group feedback on curbside recycling', *Environment and Behavior*, **27** (2), 233–50.

Dwyer, W.O., F.C. Leeming, M.K. Cobern, B.E. Porter and J.M. Jackson (1993), 'Critical review of behavioral interventions to preserve the environment: research since 1980', *Environment and Behavior*, **25** (3), 275–321.

Gardner, G.T. and P.C. Stern (2002), *Environmental Problems and Human Behavior*, Boston, MA: Allyn & Bacon.

Geller, E.S. (2002), 'The challenge of increasing proenvironment behavior', in R.G. Bechtel and A. Churchman (eds), *Handbook of Environmental Psychology*, New York: John Wiley & Sons, pp. 525–40.

Goldstein, N.J., R.B. Cialdini and V. Griskevicius (2008), 'A room with a viewpoint: using social norms to motivate environmental conservation in hotels', *Journal of Consumer Research*, **35**, 472–82.

Hogg, M.A. and G.M. Vaughan (2002), *Social Psychology*, 3rd edn, Harlow: Pearson.

Hopper, J.H. and J.M. Nielsen (1991), 'Recycling as altruistic behavior. Normative and behavioral strategies to expand participation in a community recycling program', *Environment and Behavior*, **23**, 195–220.

McKenzie-Mohr, D. (2000), 'Promoting sustainable behaviour: an introduction to community-based social marketing', *Journal of Social Issues*, **56**, 543–54.

Midden, C.J., J.E. Meter, M.H. Weenig and H.J. Zieverink (1983), 'Using feedback, reinforcement and information to reduce energy consumption in households: a field-experiment', *Journal of Economic Psychology*, **3** (1), 65–86.

Pallak, M.S. and N. Cummings (1976), 'Commitment and voluntary energy conservation', *Personality and Social Psychology Bulletin*, **2** (1), 27–31.

Schultz, P.W. (1998), 'Changing behavior with normative feedback interventions: a field experiment on curbside recycling', *Basic and Applied Psychology*, **21**, 25–36.

Schultz, P.W., S. Oskamp and T. Mainieri (1995), 'Who recycles and when? A review of personal and situational factors', *Journal of Environmental Psychology*, **15**, 105–21.

Schultz, P.W., J.M. Nolan, R.B. Cialdini, N.J. Goldstein and V. Griskevicius (2007), 'The constructive, destructive, and reconstructive power of social norms', *Psychological Science*, **18**, 429–34.

Schwartz, S. (1977), 'Normative influences on altruism', *Advances in Experimental Social Psychology*, **10**, 222–79.

Siero, F.W., A. Bakker, G.B. Dekker and M.T.C. Van den Burg (1996), 'Changing organizational energy consumption behaviour through comparative feedback', *Journal of Environmental Psychology*, **16**, 235–46.

Siero, S., M. Boon, G. Kok and F.W. Siero (1989), 'Modification of driving behavior in a large transport organization: a field experiment', *Journal of Applied Psychology*, **74**, 417–23.

Spaargaren, G. (2003), 'Sustainable consumption: a theoretical and environmental policy perspective', *Society and Natural Resources*, **16**, 687–701.

Staats, H., E. Van Leeuwen and A. Wit (2000), 'A longitudinal study of informational interventions to save energy in an office building', *Journal of Applied Behavior Analysis*, **33**, 101–4.

Staats, H., P. Harland and H.A.M. Wilke (2004), 'Effecting durable change. A team approach to improve environmental behavior in the household', *Environment and Behavior*, **36**, 341–67.

Stern, P.C. and G.T. Gardner (1981), 'Psychological research and energy policy', *American Psychologist*, **36**, 329–42.

Tajfel, H. and J.C. Turner (1986), 'The social identity theory of intergroup behaviour', in S. Worchel and W.G. Austin (eds), *Psychology of Intergroup Relations*, Chicago, IL: Nelson-Hall, pp. 7–24.

Terry, D.J., M.A. Hogg and K.M. White (1999), 'The theory of planned behaviour: self-identity, social identity and group norms', *British Journal of Social Psychology*, **38**, 225–44.

Wang, T.H. and R.D. Katzev (1990), 'Group commitment and resource conservation: two field experiments on promoting recycling', *Journal of Applied Social Psychology*, **20**, 265–75.

Weenig, M.W.H. and C.J.H. Midden (1991), 'Communication network influences on information diffusion and persuasion', *Journal of Personality and Social Psychology*, **61**, 734–42.

Wood, J.V. (1996), 'What is social comparison and how should we study it?', *Personality and Social Psychology Bulletin*, **22**, 520–37.

PART II

Challenges for local level climate change policy and alternative models for low carbon community governance

4. Transforming the nation-state through environmentalism: political influences on a multi-level governance framework in the UK

Shane Fudge

INTRODUCTION

Environmental issues have become more central to government decision-making in the UK in recent years, gaining a noticeably higher profile in relation to the traditional hierarchy of policy. The effects of industrial oil spills, ozone depletion, atmospheric pollution, and more recently the risks posed by global climate change, have all served to inform the political agenda in a much more visible way than was previously the case. A series of Energy White Papers during the last decade have been amongst the higher profile policy statements regarding the UK Government's intention to address the urgency of climate change through mainstream policy. This intent was confirmed in 2006 by the appearance of the government financed *Stern Review* which provided a comprehensive economic, social and environmental hypothesis of future scenarios should the UK fail to react to the reality of climate change. Even more significantly, the 2008 Climate Change Bill – incorporating the UK Government's pledge to cut CO_2 emissions by 80 per cent by 2050 – clearly indicates a more central role for environmental issues in consideration of the UK's present and future policy agenda.

In a broader political sense, as Giddens (2000) has argued, the last three decades have seen decision-making competencies in the West move away from a previous emphasis on top-down implementation. He points out that there has been an observable shift towards greater de-centralization, wider political engagement and more open democratic discussion on the most effective ways to implement policy in order to accommodate a changing world. Giddens suggests that the environment has been one of a number of issues to have influenced a process of 'institutional reflexivity' in the UK (Giddens, 1991) over the last three decades. Climate change, for instance, is now governed in a quite different way from the UK's previous 'regulation from the inside' position on environmental policy and its growing profile means that it also has an influence on wider policy-making processes.

The *UK Low Carbon Transition Plan* (DECC, 2009), for instance, sets out the importance of community engagement and local level decision-making in addressing climate change, but this has arguably come about as a result of a variety of external and internal forces which have been instrumental in transforming state influence on the environment in the UK. As argued above, this chapter suggests that the period which stretched from the post-war period leading up to the early 1980s was characterized by a top-down approach based upon a 'national' style of implementation. It is argued that this then became a more complex political structure, whereby decision-making processes were opened up to orchestration through multiple points of engagement – primarily through deepening European integration and increasing international political influence. The final section of the chapter considers how the more recent influence of local government and community-led responses to the climate change debate in the UK can also be traced in part to the long-term impact of these processes, where they continue to push for further institutional shifts in national policy-making. It is suggested that Agenda 21, for instance, began a political dialogue on the significance of local action in guiding sustainable development initiatives. As the chapter points out, the UK's more recent targets in relation to reducing CO_2 emissions quite clearly acknowledge that the greater emphasis on targeting individuals and behaviours at a grassroots level can only be fully realized by incorporating and encouraging local government and community-led action as a more central aspect of the political process.

UK ENVIRONMENTAL POLICY IN THE POST-WAR PERIOD

> It was characteristic of organized capitalism that a whole range of economic and social problems were thought to be soluble at the level of the nation-state. Issues of poverty, health and the environment were to be dealt with through national policies, especially through a Keynesian welfare state which could identify and respond to what one might call the risks of organized capitalism. (Lash and Urry, 1994, p. 292)

Since the end of the Second World War, environmental policy in the UK has evolved in relation to a number of interlinked issues, notably: the world's increasing economic and political interconnectedness and heightened awareness of the trans-boundary consequences of environmental problems; intensified media and literary coverage of environmental issues; and increased environmental interest group activism alongside their deeper engagement in challenging more formal political processes.

The ethos behind environmental policy during the period leading up to the early 1980s was characteristic of the political rationale which lay behind the vertical style of 'command and control'-style regulation favoured by British Government policy implementation. McCormick (1995, p. 159) makes the point that 'British policy makers have traditionally viewed themselves as custodians of the public interest, and have felt sure they could understand this interest with minimal reference to the public itself'. Regulation on issues such as pollution control, for instance, was often implemented through a system of 'voluntary compliance', where according to Beck (1999), scientific knowledge and expertise were primarily aimed at ensuring the efficient production of 'goods' as opposed to the 'bads' that later became apparent. Beck suggests that any awareness of the possibility of negative environmental impacts during this period was largely subsumed by a dominant set of ideas which implied that: a) any such impacts could ultimately be controlled by a top-down regulatory system of expertise; b) environmental impacts that could be discerned would be contained within the parameters of nation-state control; and c) unforeseen environmental impacts would ultimately be outweighed by the perceived benefits of industrial modernization and could therefore be justified under the banner of social and economic progress.

Lowe and Warde (1998) have listed a typology of the distinguishing features that characterized this 'regulation from the inside' which then set the agenda for the dominant characteristics of environmental policy in the UK up until approximately the early 1980s. They reason that, typically, environmental policy during this period was characterized by:

- *Low politics*: as argued above, environmentalism in this period was not considered to be a crucial component or consideration in the drive towards facilitating UK economic performance. Environmental policies were not a primary concern of the principal political parties during this time and were not drivers of mainstream government decision-making. During this period, environmental policies were delegated primarily to devolved administration units, principally to the Department of the Environment (DoE).
- *Devolved fragmentation and local implementation*: organization and implementation of environmental policy during this period tended to be the responsibility of local authorities, quangos and

semi-independent inspectorates which provided 'technical expertise' from what were relatively de-politicized settings. The delegation of environmental responsibilities to local authorities and the range of policies with which they were charged were based upon the belief that environmental concerns were best dealt with at the local level.

- *Disjointed incrementalism*: environmental policy during the post-war period was typified by a lack of overall coherence. The DoE itself acknowledged in 1976 that its approach to environmental policy was essentially pragmatic, pointing out that 'as an early industrial country, Britain has generally built up her law and administration stage by stage in response to particular problems' (DoE, 1976, p. 6). The result of this incremental and primarily reactive policy approach to the environment was the clear absence of overall aims through which to lead a more coordinated policy agenda. Concurring with this observation, McCormick (1991, p. 10) points out that British environmental policy has generally been 'prone to the kind of *ad hoc* improvisation and piecemeal responses that characterized the policy process of this time in general'. He points out that the consequences of this were that UK environmental policy in this era was often typified by a lack of clear direction and an often confusing framework of institutions and laws.

- *Reliance on scientific and technical expertise*: the pragmatic problem-solving approach adopted during the post-war period regarding the management and regulation of environmental policies left considerable scope to the judgement of particular individuals in interpreting and implementing policy. In this way, legislation often tended towards being unclear in relation to the realization of targets and standards of policy execution. Policy development was also heavily reliant upon the 'expert consensus' approach achieved through the work of advisory committees, commissions and quangos.

- *Informal regulation*: Vogel (1983) suggests that environmental policy during this period was primarily based upon an informal approach rather than being based upon a framework of law. He points out that this often led to a situation where there was considerable leeway in the drawing up of policy regulations. Furthermore, the system of self-monitoring within a particular policy network worked primarily through a system of 'negotiated compliance'.

- *Close consultation with affected interests*: the policy system during this period was characterized by a particular technical approach and an aim towards consensus. This therefore meant that the regulation and implementation of environmental policy in the UK often took place primarily with the economic and producer interests that it was intended to influence. While environmental groups, such as those with specialist expertise in nature conservation, landscape protection, historic conservation, and town and country planning, had relatively easy access to the DoE and its subsidiary departments, they were invariably excluded from other major government departments which were making decisions with what were later found to be profound environmental implications, including Transport, Trade and Industry, Agriculture and the Treasury (Lowe and Ward, 1998, pp. 7–8). Thus, environmental policies at this time could generally be considered to constitute a low political agenda where their contribution to the 'high politics' of national economic performance and management – particularly in terms of regulating UK energy markets and institutions – was at best uneven.

As McCormick (1995, p. 160) argues, pollution was generally not seen as a problem during this period. This attitude was particularly illustrative in the coal-mining areas and industrial heartlands, where a reluctance 'to accept less pollutive, alternative energy sources such as natural gas' was also linked to the prioritization of other issues such as ensuring full employment. There was a particular fear among politicians and business leaders that increased environmental controls and regulations would have the effect of destabilizing the existing manufacturing sector where more punitive measures could potentially discourage the establishment or relocation of new industries.

THE STOCKHOLM CONFERENCE, THE ENVIRONMENTAL MOVEMENT AND THE INFLUENCE OF EUROPEAN ENVIRONMENTAL LEGISLATION

Building on the concerns expressed by the environmental movement in the 1960s, and also in the 1972 publication of *The Limits to Growth*, the 1972 UN Conference on the Human Environment in Stockholm can be viewed as a landmark change in the way in which a wider political awareness began to emerge regarding the importance of political regulation in promoting environmental aims in policy. The Stockholm Conference was significant in that it triggered three particular institutional changes which would begin to challenge the ways in which Western governments would address environmental issues in the future. First, it signified the growing importance of recognizing that polluting activities and environmental degradation should now become very real concerns at a formal political level;

secondly, it specified that that political framework of the debate would be globally defined as accumulating evidence pointed to the transnational status of environmental issues; and thirdly, it set in place an agenda for further action and policy precedents that would be taken in the future of environmental policy. The formation of the United Nations Environment Programme at this time formalized some of the principal objectives of the Stockholm Conference, particularly the aim to institutionalize greater partnership working and international problem-solving collaboration.

European Integration and the Environment

The European Community was represented at the Stockholm Conference by the then membership of six nations: a dialogue which began to set the scene for a more substantial, interstate environmental policy at European level. Up to this point, Community-instigated environmental policy was fairly marginal, and Lowe and Ward assert that 'the 1957 Treaty of Rome that established the EEC made no mention of environmental protection'. Up until the Stockholm Conference, they suggest that the primary concern of European leaders was with the effective construction of a free trade area through which to 'reflect the dominant concerns of post-war Europe with economic reconstruction, modernization and improved living standards' (Lowe and Ward, 1998, p. 11).

Jordan (2006) argues that the above developments were instrumental in beginning to challenge the insular nature of environmental policy in the UK. He suggests a number of reasons, for instance, why the development of European level environmental policy began to exert a more transnational influence over the UK's policy structure of the time. He argues that the Europeanization of the policy arena since the UK joined the then Common Market in 1973 had the effect of: influencing UK environmental policy to move into a more open, transparent and democratic policy structure, where it has empowered and aided greater lobbying power for UK environmental groups and organizations; increased scientific-monitoring and technical-processing methods on the condition of the environment; provided an effective forum for sharing national environmental best practice through which the UK can then benchmark its own practices; provided a consistent, external set of laws and institutions through which to monitor the UK's environmental progress or otherwise; and enabled the other member states of the EU to obtain a better idea of the ethos and traditions behind environmental policy in the UK (Jordan, 2006, pp. 13–14).

Jordan argues that the ongoing development of European environmental policy has often worked to 'lock in' environmental initiatives in the UK and also in the other member states, where the benchmarks for policy have often been set by countries which have been the most forward thinking with regard to their own environmental policies. A good illustration of this occurred in the 1980s in a case of 'laggards versus leaders' in the European Community when Germany and the Netherlands were both able to successfully challenge the UK over its sulphur pollution levels, which had affected forestry and woodland areas in both these countries. As a part of the European Community, the UK was forced to revise its environmental standards in this domain and the status of national legislation over acid rain pollution was challenged through the European Commission and the European Court of Justice during the 1980s. Jordan has suggested that while the growing influence of international treaties pertaining to environmental issues had not always been binding – based as they were primarily on intergovernmental cooperation – the development of an effective EU environmental policy has been instrumental in binding many of the issues and agreements that had been made into enforceable directives. The 1987 Montreal Protocol and the 1992 Climate Change and Biodiversity Conventions, for instance, have been instrumental, as Jordan has pointed out, 'in giving a strong push to international efforts to tackle ozone depletion, climate change and biodiversity loss respectively' (Jordan, 2006, p. 11).

Interest Groups and the Environment

As Jordan (2006) argues, one of the most important ways in which UK environmental policy has been opened up by the growing political profile of environmental issues has been the way in which nongovernmental organizations (NGOs), civil society organizations and even the general public have been able to become more active in influencing policy and political agendas. As Beck (1999) has argued, this has been a more obvious example of the increasing influence of what he has termed 'sub-politics', whereby the more traditional top–down political structures regarding the governance and management of contemporary risk have been undermined and influenced by what he terms 'reflexive modernization'. Reflexive modernization, argues Beck, is the process whereby the influential relationship between science and modernity has been opened up to challenge by the more unforeseen by-products of industrialization – particularly environmental degradation. He suggests that the increase in grassroots level activism has been indicative of the democratic drive towards what he describes as an ethical globalization (Beck, 1999, p. 2). In the UK in particular, many of the groups and organizations which had found themselves marginalized by the insular structure and institutional design of UK environmental policy became drawn to Brussels as an increasingly influential political structure through which to circumvent national government regulation and the often depoliticized risks of this period. As Jordan (2006) points out, over time this area of policy has become particularly illustrative of a domain where the EU has been able to assume the role of an arbiter or 'higher authority' for various environmental interests, mediating between national and supranational tiers of decision-making, and also incorporating environmental groups and organizations outside of the 'formal' political sphere.

The extent to which NGOs actually influence 'high political' issues in this way continues to be debated in the literature (Grant, 1993; Greenwood, 2003). However, there is no doubt that their presence at high level environmental conferences, such as at the annual United Nations Framework Convention on Climate Change (UNFCCC) Conference of the Parties, provides for continued monitoring of action on climate change from the third sector, whether at national or international level.

The above developments were all instrumental factors in highlighting and then challenging the UK's post-war position on environmental policy. The particular limitations that were exposed by the growing influence of international/EC policy were that:

- it was a mistake to assume that environmental issues could continue to occupy the status of 'low politics';
- environmental policies could no longer be constituted as 'depoliticized' issues which could be isolated from their particular circumstances by independent 'experts';
- there needed to be a greater emphasis on proactive rather than reactive environmental policies alongside greater overall coherence and joined-up strategies;
- there should now be a consideration of real targets and/or guidelines to direct environmental policy initiatives;
- there needed to be a more formal framework based on legal principles as opposed to informal regulation;
- there needed to be close consultation with affected interests where 'high' political areas such as economic performance would invariably hold sway.

THE POLITICS OF CLIMATE CHANGE

Good environmental management is not a technical exercise separate from everyday economic and political life, or something tacked on after development. It can only come about when environmental values are embedded within economic and political systems. For this to occur, it is first necessary to understand something of those systems and the basic tensions that affect them. In modern political systems, particularly democratic ones, an enduring tension exists between top-down forces and bottom-up aspirations. (Carley and Christie, 2000, p. 121)

More than any other environmental issue, the urgency of climate change increasingly shapes the policy agendas of national governments. While the globalization of the world economy still provides the most vivid illustration of what Ohmae (1995) calls the arrival of the 'borderless state', the increasing consensus on the science of climate change has meant that no government can stand outside of its environmental and political implications. The urgency of climate change for instance has had a noticeable impact on policy-making and governance in the UK, where it has:

- led environmentalism into mainstream policy-making;
- fostered negotiation between ever-increasing numbers of different interests;
- introduced a form of risk typified by non-localized status, meaning that policy initiatives must be implemented to varying degrees at national, international *and* local levels.

Finally, uncertainties over climate change science have meant that initiatives remain characterized by an ongoing search for the most effective policies and accompanying governing frameworks.

The increasing political significance of the UNFCCC provides a graphic illustration of how the world political landscape is increasingly organizing at a trans-national level around the ownership and threat of environment and resource issues. Indeed, the growing threat of climate change was a major factor in the negotiations which culminated in the Rio Declaration on Environment and Development: an international treaty which itself gave birth to the UNFCCC. Following on from the 1989 Toronto Conference on Sustainable Development and the 1990 meeting of the International Panel on Climate Change, the 1992 Rio Declaration on Environment and Development resulted in several key policy developments, including 27 principles on the goal of establishing a new and equitable global partnership. As Irwin (2001, p. 42) has argued, the development of Agenda 21 represented a 'governing framework' within which national governments could now operate, in order to better achieve more environmentally and socially sustainable conditions. Perhaps the most important principle introduced by Agenda 21 was that 'social equity and wide public participation would be central to this framework' (Irwin, 2001, p. 42).

Agenda 21 was instrumental in highlighting the continued need for national governments to reappraise the political level through which policies aimed at promoting sustainable development would prove to be most effective. It could be argued in fact that the programme that was developed out of Agenda 21 – Local Agenda 21 (LA 21) – provided a new impetus to the mobilization of a more local response to sustainable development as a complement to global initiatives that were being developed. Echoing some of the aims characteristic of the EU's own 'subsidiarity principle' – agreed at Maastricht a year earlier – the primary aim of LA 21 dictates that local governments should now consult more directly with the key stakeholders in their area in order to reach consensus on drawing up long-term environmental action plans in each area. An acknowledgement that top-down policy could only go so far – particularly with regard to the increasingly complex questions that were being asked of political leaders in constructing effective policies to address climate change – one of the aims behind LA 21 was to encourage projects which could be 'embedded' within bottom–up social, cultural and economic particularities. It was argued that these would hold the potential to be more effective than top–down solutions in enabling citizens and individuals to recognize their own roles in contributing to more sustainable ways of living. The ethos behind LA 21 was also drawn up with the aim of encouraging greater stakeholder engagement in the development and process of policy by providing citizens with the means to engage more fully in the wider political debate on sustainable development and the process of governance.

Agyeman and Evans (1994) have suggested that LA 21 exhibits aspects of what many regard as some of the key issues central to enabling sustainability to come about as a policy goal. As they point out, 'these issues now include community environmental education; democratization; balanced partnerships between public and private sectors; and integrated policy-making' (1994, p. 153). Indeed, Barry (1999, p. 153) argues that one of the most important functions of LA 21 has been its potential as a political catalyst in facilitating a process of 'democratic ecological governance'. Driven by the increased emphasis on stakeholder engagement in policy, he argues that:

> Local Agenda 21 can be invaluable in informing policy with . . . 'a modern commons type regime' which can secure an ecologically rational form of collective ecological management. Making people aware of the interconnectedness of human wellbeing (including economic considerations) can give them a greater say in formulating local government policy, does highlight the connection between long-term human self-interest and environmental responsibility. (Barry, 1999, p. 154)

LA 21 AND THE UK: DEMOCRATIZING CLIMATE CHANGE?

The UK Government began to take an active interest in climate change from around the time of the late 1980s. The appearance of *This Common Inheritance* in 1990 has been held by some to be an endorsement of the then Thatcher Government viewpoint that climate change was now one of the biggest environmental challenges facing the planet and the onus was now on the present generation to take responsibility by taking action to reduce greenhouse gas emissions. There were even UK targets (unrealistic as it turned out) to keep greenhouse gas emissions within their present levels during this period in the face of scientific claims that emission levels were rising.

Carley and Christie (2000, p. 212) have pointed out that the election of a Labour Government in the UK in 1997 suggested that a greater commitment to LA 21 would tie in well with ideas regarding a new political settlement that was being constructed between state intervention and market primacy: a settlement which seemed to support a more creative approach to policy-making. The rhetoric from the new regime also suggested that there would be greater support from policy makers for environmental and social initiatives than may previously have been the case. In contrast to the highly centralist approach deployed by the previous Conservative administration, the Labour Government itself seemed to support the growing awareness in policy circles that top–down administrative approaches would only go some way towards bridging the intricate relationships between citizens, institutions, issues and policy delivery mechanisms.

This perceived willingness to devolve some aspects of policy implementation and decision-making away from formal political institutions was thrown into sharp focus by the growing complexities of how to incorporate sustainable development in policy. Climate change itself was not a priority for the Labour Government when it came to office in 1997 (Helm, 2004). A structural shift in energy use during the 1990s' 'dash for gas' – an unforeseen 'windfall' from energy privatization – had seen the UK's greenhouse gas emissions begin to tail off. This enabled the new government to confidently sign up to the Kyoto Protocol in 1997 and at the same time to commit to its own national target of a 20 per cent reduction in Greenhouse Gas emissions by 2010, according to 1990 baseline levels.

However, this viewpoint came open to challenge as the UK's emissions began to increase again during the mid-1990s, mostly due to growth in road transport and air travel (Royal Commission, 2000). There was also a growing awareness that the energy demand in housing accounted for as much as 40 per cent of the UK's CO_2 emissions total (Jones et al.,

2000). Critics pointed out that this was evidence that policy initiatives would now have to be constructed in order to unlock more direct patterns of consumption, such as those associated with travel, eating habits, leisure practices and living patterns, if the government were to be serious about reducing the UK's levels of carbon emissions. Therefore, while the implementation of LA 21 in the UK has been somewhat uneven (Fudge and Peters, 2009), the practical difficulties of reaching UK targets on CO_2 emissions have encouraged policy makers to continue to work with and to encourage more 'bottom-up' initiatives in order both to engage communities and to persuade individuals to live and work in more environmentally and socially sustainable ways. The 2009 UK Energy White Paper has highlighted the importance of trying to enable local knowledge of this kind to become better integrated into more mainstream policy approaches on reaching CO_2 emissions targets. For instance, it is argued that local authority initiatives could provide better political leverage in encouraging individuals to recognize their own role in contributing to more sustainable levels of energy consumption, particularly in households and in transport practices (DECC, 2009).

CONCLUSION: ADDRESSING CLIMATE CHANGE THROUGH MULTI-LEVEL GOVERNANCE

This chapter has explored the development of environmental policy in the UK from the post-war period, arguing that it has changed considerably over the last fifty years due to a variety of influences which have acted upon and changed the conception of nation-state politics during this time. It was argued that environmental policy in the UK originally occupied the status of 'low politics'; was overseen on a de-politicized level by groups of independent experts; lacked overall coherence and tended to be reactive rather than proactive; there were no real targets or legal standards to guide policy; and what policies there were tended to take place in close consultation with affected interests where 'high' political areas, such as economic and business concerns, would invariably hold sway. It was argued that these factors began to change, primarily in relation to the increasing influence of transnational economic and political cooperation in environment and sustainability issues at the global level. It was suggested that changes in the nature of democracy could also be traced to the emergence of other actors outside of formal political platforms – notably NGOs and interest organizations – which were also instrumental in raising questions related to environmental issues being dealt with solely by politicians and groups of 'experts'. In addition, membership of the European Community began to have a big influence on member states' policies; an example being the way in which 'green leaders' during the 1980s in Sweden and Germany were able to exert pressure on the UK as an environmental laggard – politically through the European Commission and then legally through the European Court of Justice.

Largely as a consequence of how environmental policy has been 'primed' by these changes, it could be argued that climate change policy in the UK now exemplifies what Hooghe and Marks (2001) have described as a 'multi-level governance' policy framework. As argued above, the global politics surrounding climate change have more recently had a particularly noticeable effect on encouraging a recognizable sub-national level of governance in the UK which is becoming more active within a framework of national policy measures and targets. As the chapter has argued, the evolution of environmental policy into a global concern has meant that national policy initiatives can no longer stand in isolation but are housed within the context of European legislation, international agreements and, increasingly, local influence. Thus, the national level is now arguably more a catalyst than an agenda setter: particularly noticeable in an area such as climate change. In the UK, for instance, the government has set in place a framework for the debate – in this case its own national targets on pollution which are embedded in law. However, due to the influence of the issues addressed in this chapter, the successful implementation of policy will now invariably be predicated upon its successful coordination across different levels of governance in order to achieve a common goal.

REFERENCES

Agyeman, J. and B. Evans (1994), *Local Environmental Strategies*, Harlow: Longman.

Barry, J. (1999), *Environment and Social Theory*, London: Routledge.

Beck, U. (1999), *World Risk Society*, Cambridge: Polity Press.

Carley, M. and I. Christie (2000), *Managing Sustainable Development* (2nd edn), London: Earthscan.

DECC (Department for Energy and Climate Change) (2009), *The UK Low Carbon Transition Plan: National Strategy for Climate and Energy*, London: DECC, and Norwich: The Stationery Office, accessed at www.decc.gov.uk/en/content/cms/publications/lc_trans_plan/lc_trans_plan.aspx.

DoE (Department of the Environment) (1976), *Pollution Control in Britain: How it Works*, London: HMSO.

DoE (1990), *This Common Inheritance: Britain's Environmental Strategy*, London: HMSO.

Fudge, S. and M. Peters (2009), 'Motivating carbon reduction in the UK: the role of local government as an agent of social change', *Journal of Integrative Environmental Sciences*, **6** (2), 103–20.

Giddens, A. (1991), *Modernity and Self Identity: Self and Society in the Late Modern Age*, Cambridge, UK and Boston, MA: Polity Press.

Giddens, A. (2000), *The Third Way and its Critics*, Cambridge, UK and Boston, MA: Polity Press.

Grant, W. (1993), 'Pressure groups and the European Community: an overview', in S. Mazey and J. Richardson (eds), *Lobbying in the European Community*, Oxford: Oxford University Press, pp. 27–46.

Greenwood, J. (2003), *Interest Representation in the European Union*, London: Palgrave.

Helm, D. (2004), *Energy, State and the Market*, Oxford: Oxford University Press.

Hooghe, L. and G. Marks (2001), *Multi-Level Governance and European Integration*, Lanham, MD: Rowman and Littlefield.

Irwin, A. (2001), *Sociology and the Environment*, Cambridge, UK: Polity Press.

Jones, E., M. Leach and J. Wade (2000), 'Local policies for DSM: the UK's Home Energy Conservation Act', *Energy Policy*, **28**, 201–11.

Jordan, A. (2006), *The Environmental Case for Europe: Britain's European Environmental Policy*, CSERGE working paper EDM 2006–11, Norwich: University of East Anglia.

Lash, S. and J. Urry (1994), *Economies of Signs and Space*, London: Sage.

Lowe, P. and S. Warde (1998), 'Britain in Europe: themes and issues in national environmental policy', in P. Lowe and S. Warde (eds), *British Environmental Policy and Europe: Policy and Politics in Transition*, London: Routledge, pp. 3–30.

McCormick, J. (1991), *British Politics and the Environment*, London: Earthscan.

McCormick, J. (1995), *The Global Environmental Movement* (2nd edn), London: John Wiley and Sons.

Meadows, D.H., D.L. Meadows, J. Randers and W.W. Behrens (1972), *The Limits to Growth*, London: Pan Books.

Ohmae, K. (1995), *The End of the Nation State: The Rise of Regional Economics*, London: Harper Collins.

RCEP (Royal Commission on Environmental Pollution) (2000), *Report on Environmental Pollution*, London: The Royal Commission on Environmental Pollution, accessed at www.rcep.org.uk.

Stern, N. (2006), *The Economics of Climate Change: The Stern Review*, Cambridge: Cambridge University Press.

Vogel, D. (1983), 'Comparing policy styles: environmental protection in the US and Britain', *Public Administration Bulletin*, **42**, 65–78.

5. The role of local authorities in galvanizing action to tackle climate change: a practitioner's perspective

Simon Roberts

INTRODUCTION

The role of local authorities in tackling climate change is increasingly identified as a key component of a concerted national effort to curb carbon emissions. In the UK, new performance indicators have been put in place such that local authorities will start to be measured and performance managed on how well they use their influence to curb emissions locally.

This chapter explores the potential extent of that influence (examining also reasons for its limitations), the necessary conditions for successful action by local authorities (which are currently relatively rare), and the relative merits of the range of approaches being used in an attempt to establish such conditions. It offers a practitioner's perspective; the Centre for Sustainable Energy (CSE) is one of the UK's leading energy charities which has been working closely with local authorities on initiatives to tackle climate change for several decades and more recently has led key research studies for the UK Government and the Local Government Association into the role of local authorities.

THE OFFICIAL RISE OF LOCAL AUTHORITIES AS 'KEY' PLAYERS IN TACKLING CLIMATE CHANGE

At the turn of the twenty-first century, the UK Government's Climate Change Programme (UKCCP) laid out for the first time its suite of policies designed to deliver the UK's international and national commitments to reducing greenhouse gases, and in particular carbon dioxide emissions (HM Government, 2000). The UKCCP expressly acknowledged the 'unique and critical role' that local authorities can play in ensuring delivery of these largely nationally driven policies to curb carbon emissions at a local level.

Three years later, the Energy White Paper of 2003 talked of local authorities as having a 'key role' in delivering national carbon emission reduction targets (HM Government, 2003). At first sight this endorsement appeared to demonstrate an appreciation of the importance of local action on carbon emissions and the role of local authorities within such action.

Yet neither of these documents was clear about the precise nature of this 'key role' for local authorities. They indicated no sense of how well it was assumed the role was being, or would be, fulfilled by local authorities. And there was certainly no indication of the impact on UK carbon emissions which would be achieved by widespread and high quality fulfilment of the role. As the UK Government started to acknowledge a potentially significant shortfall between the forecast impact of its policies and the targets it was trying to achieve, it launched a Review of the UKCCP in 2004 (HM Government, 2004).

Driven perhaps by this shortfall to a desperate search for 'every tonne of carbon saving we can get from anywhere that can deliver it',[1] a study of the potential of local and regional bodies to influence reductions in carbon emissions was jointly commissioned by the UK Department for the Environment, Food and Rural Affairs (Defra) and Department of Trade and Industry (DTI). The study undertaken by a team led by the CSE (CSE et al., 2005), was specifically targeted at finding evidence of the impact on local carbon emissions of actions by local authorities and regional public bodies (specifically Regional Development Agencies and Regional Assemblies).

However, the study found no such evidence or indications of any systematic attempts to compile it. Even where 'best practice' case studies, of an individual local authority's work on, say, improving its housing stock, had been written up, they rarely contained detailed quantitative analysis of the carbon savings delivered and the specific contribution of the local authority to the achievement and any associated costs.

The study also found no structured description of the role of local authorities in delivering carbon reductions in their localities. More specifically, it was unable to unearth any assessment of how the role of local authorities fitted within national (and, indeed, international) efforts to curb carbon emissions; there was no clearly stated understanding of the relationship between local, regional and national action.

No one seemed to have answered the question of what local authorities 'could do' or 'were for' in the context of tackling climate change. As a result, there appeared at the time to be two schools of thought in the debate regarding local authorities and action on climate change. One was dominated by the more active local authority officials who tended to assert the pre-eminence of local action with calls for powers to empower local agencies to do more. This particular argument seemed to be driven by a belief that local authorities needed to play a more prominent role in order to make up for a perceived failure of national policies and programmes on climate change.

The other school of thought – prominent among UK government officials – tended towards seeing local action as, at best, a 'nice to have' option and, at worst, a messy and unreliable substitute for effective, nationally determined programmes. Interestingly, despite having authored both the UKCCP and the 2003 Energy White Paper, advocates of this perspective in practice gave little credence to the idea that local authorities might actually have a key role in national efforts to cut carbon emissions.

In order to address this apparent lack of understanding, Defra and DTI commissioned a team led by the CSE to define more clearly the role which local authorities could play in cutting carbon emissions (CSE et al., 2005). This involved: (a) examining the nature of the problem of carbon emissions; (b) considering the roles and responsibilities already accorded to local authorities in other aspects of policy and service delivery; and (c) exploring the types of political and practical interventions available.

Through a consideration of the findings of this research – described in more detail below – and an accompanying assessment of how poorly local authorities were generally performing in this role, the UK Government was persuaded in 2006 that action to improve the performance of local authorities in relation to cutting carbon emissions would make a meaningful contribution to the revised UKCCP.

As a consequence, when the updated UKCCP was published in 2006, the Government announced that it now wanted to see 'a significant increase in the level of engagement by local government in climate change issues' (HM Government, 2006, p. 106). Specifically, the UK Government undertook to introduce into the 'performance assessment system for local authorities' a performance indicator which focused on their achievements in stimulating reductions in carbon emissions caused by energy and non-motorway transport use in local authority areas.

This performance indicator, or 'NI 186', was introduced in England in April 2008 and it proved to be one of the most eagerly adopted targets for local authorities as they worked with their Local Strategic Partnerships to identify local priorities. All local authorities will now be measured and their performance managed on how well they use their influence to curb emissions within their political jurisdiction. Those local authorities which have adopted this indicator as one of their core priorities were required to set targets for reducing carbon emissions and will be additionally assessed on their progress towards this target.

The status of local authorities in delivering a low carbon future continues to grow in more recent Government policy statements. In the 2009 *UK Low Carbon Transition Plan* (HM Government, 2009), for instance, local authorities, and their role as a 'vanguard' of local and community action on climate change, are mentioned 33 times. As the document itself explains:

> The Government wants to encourage and empower local authorities to take additional action in tackling climate change, where they wish to do so. It believes that people should increasingly be able to look to their local authority not only to provide established services, but also to co-ordinate, tailor and drive the development of a low carbon economy in their area. (HM Government, 2009, p. 94)

From a rather ill-defined and half-hearted endorsement of this agenda, the *UK Low Carbon Transition Plan* suggests that the role and purpose of local authorities in tackling climate change have now become much more central to national government thinking on climate change strategy in the UK. This change appears no longer to be motivated by the earlier viewpoint that 'we-need-everything-we-can-get' but by a genuine understanding of how effective action by local authorities is a vital and necessary condition for success in national efforts to meet the ever-tightening targets for carbon emission reductions.

DEFINING THE ROLE OF LOCAL AUTHORITIES IN TACKLING CLIMATE CHANGE

As outlined by research carried out by the CSE in both 2005 and 2007, the role of local authorities in tackling climate change falls naturally from the way in which virtually every facet of their existing activities, roles and responsibilities will effect some level of impact on carbon emissions – either directly or through the influence or control they exert on another person or organization. In the majority of cases, it is only local authorities which carry out these activities. It also derives naturally from the nature of commonly proposed approaches to tackling climate change.

That these possibilities were not recognized much earlier by the Government may have derived from the way in which climate change policy in the UK was initially addressed in policy as other 'pollution' problems had been – with a clear line of causation (the burning of fossil fuels) and therefore a clear focus for policy intervention (reduce the use of fossil fuels). Such a formulation ignores the extent to which the identified cause (the burning of fossil fuels) is embedded so deeply in all aspects of economic and social life; it therefore tends to over-simplify the nature, and under-estimate the scale, of policy response required. It is perhaps only in the last two or three years that the Government has demonstrated a more complete understanding of the full extent of economic and social changes implied by an adequate response to the threat of climate change.

It has already been suggested that the role and activities of local authorities mean that they are ideally placed to tackle some of the complexities of climate change. Therefore, it is worth reviewing the extent to which this latent capacity to influence carbon emission reductions is already embedded across the responsibilities of each local authority. It should be noted that this capacity to influence exists whether or not it is being acknowledged by a local authority or being applied to achieve positive effect. The capacity of local authorities to influence climate change policy in the UK derives from:

- the services which they already deliver;
- the strategic roles they play;
- the regulatory influence they have to enforce national standards and directives; and
- the relationship (and therefore potential influence) they have with the citizenry, voluntary and business sector and/or other public bodies in their vicinity.

As a result, local authorities have:

- direct connections with individual households, community groups and businesses by virtue of existing service provision and electoral relationships;
- opportunities to identify, bring together and support local organizations and encourage businesses to provide services which reflect local need and circumstance;
- potentially strong ability to establish and maintain a sense of local identity and civic pride which can make national and global issues seem locally relevant; and
- a democratically accountable role to provide civic leadership.

To look at it another way, most of the day-to-day experiences of public service and civic leadership 'enjoyed' by households, businesses and other organizations are principally mediated by local authorities.

More specifically, and taking account of existing structures and responsibilities,[2] local authorities are already expected to:

- manage their own buildings, housing stock and staff activities and procure equipment and a wide range of services;
- deliver a range of services to the public (housing, education, social services, waste, leisure/tourism, culture, etc.);
- establish and control the local planning strategy;
- coordinate local regeneration and economic development activity;
- manage and/or influence public sector investment in local infrastructure;
- enforce building regulations and trading standards;
- provide civic leadership within their communities, encouraging behavioural change and leading by example (for example, through Local Strategic Partnerships and Local Area Agreements);
- create and support effective partnerships (with each other and across sectors) to meet defined objectives;
- make nationally significant issues locally relevant and motivating;
- promote community well-being; and
- showcase good practice.

Therefore, a local authority is often in an ideal position to influence positive action on reducing emissions principally through active engagement across all of these existing activities.

Research carried out by CSE (CSE et al., 2005; CSE, 2006, 2007) provides a more detailed exploration of how to achieve such integration, in particular through the use of the Local and Regional Carbon Management Matrix. First developed by CSE, this is now widely used by local authorities themselves (see CSE, 2006). The Matrix itself outlines 49 different existing roles and responsibilities of local authorities which are themselves relevant areas of action, or 'levers', for tackling carbon emissions. It provides both a tool for assessing current performance and a guide for improving it by detailing the conduct likely to secure a 'weak', 'fair', 'good' or 'excellent' rating for each potential area for action.

UNDERSTANDING THE IMPORTANCE OF LOCAL AUTHORITY ACTION ON CARBON EMISSIONS

There are three fundamental reasons why action by local authorities on cutting local carbon emissions is a vital component of a coherent national effort to deliver carbon emission reduction objectives. First, implementation of carbon emission reduction is highly diffused – requiring a sustained change in behaviour, housing performance and consumer choices by every householder, transport user and business in the country. Secondly, amongst the individuals and groups who need to implement these changes, the current levels of motivation to act and the understanding of required actions are still relatively limited. And thirdly, the tools and technologies, services and skills to enable action are not all widely available and are currently often found in smaller organizations (voluntary, business or academic), which can fall 'below the radar' of national bodies.

These issues confirm the need for local action. They also point to a focus on changing attitudes, building understanding and motivation to act, and enabling new partnerships and service developments to test and deliver the necessary changes to support a lower carbon society. This has been reinforced in reports to Defra (Futerra, 2005; CSE with CDX, 2007) on climate change communications and individual behaviour change. Both reports concluded that there is a vital and fundamental role for local agencies in enabling these changes.

As outlined above, through their existing attributes, roles and responsibilities local authorities have the potential to provide this function. Of course, this does not mean that local authorities are currently applying these attributes and roles to the delivery of carbon emission reductions. But it is difficult to imagine a successful national effort to reduce carbon emissions or eliminate fuel poverty in which they do not.

THE LIMITATIONS OF LOCAL AUTHORITY INFLUENCE ON CARBON EMISSIONS

The fact that local authorities can have influence over carbon emissions in their localities does not necessarily mean that such influence is quantitatively significant. An analysis of the 2006 UK Climate Change Programme conducted by CSE for the Local Government Climate Change Commission (CSE, 2007) concluded that the extent of local authority influence on the success of national carbon reduction policies in cutting carbon emissions was relatively limited.

The research suggested that this was for three reasons. First, local authorities only have influence over *some* aspects of the measures outlined in the 2006 UKCCP (for example, very limited participation in the EU emissions trading scheme but significant influence in terms of area-based household energy-saving schemes).

Secondly, influence by a local authority on carbon emission reductions will also depend on how well it exerts that influence through the various roles it performs. For example, if a local authority is quite poor at engaging with community organizations; treats with indifference opportunities to secure funding from energy suppliers for energy efficiency measures in homes in its area; or fails to secure a productive relationship with local energy advice providers, it will not be an effective catalyst in reducing carbon emissions in these areas. The *quality* of a local authority's performance in its surrounding community is therefore a key factor in the scale of influence it will be able to exert.

Finally, the apparently limited influence exerted by local authorities so far in reducing carbon emissions in their communities could also be related to the fact that the Government seemed reluctant to commit in the UKCCP 2006 to measures which relied significantly on good local authority performance. This was because of both (a) a reluctance to commit to local authority action in the absence of robust data of its impact (as outlined in CSE et al., 2005) and (b) some uncertainty about how best to go about improving performance by local authorities in tackling climate change.

Overall, the assessment outlined in CSE (2007) indicated that the quality of local authority performance had an influence of approximately 7.5 per cent across the UKCCP 2006. In other words, the levels of carbon emissions from a local authority area where the local authority performs poorly are likely to be 7.5 per cent higher than an area where the authority is performing excellently. This is not the same as saying local authorities only influence 7.5 per cent of the carbon savings in the UKCCP. On the contrary, the assessment suggested that they had an influence over 45 per cent of the carbon savings delivered by UKCCP measures. If only those 'influenceable' measures are considered, the variability of local carbon emissions due to the quality of local authority performance more than doubles to 18 per cent.

As outlined above, the potential impacts on carbon emissions of effective local authority action were likely to have been under-estimated because the policies in the UKCCP 2006 were not designed with the role of local authorities in mind. The UKCCP assessment of potential carbon savings therefore missed opportunities for carbon emission reduction associated with effective local authority action. Bearing in mind the better feel for the potential of local authority impact which is now embedded within the 2009 *UK Low Carbon Transition Plan* (HM Government, 2009), it should prove useful to undertake a new analysis of the extent of local authority influence on the achievement of national carbon emission reduction targets.

ESTABLISHING THE CONDITIONS REQUIRED FOR SUCCESSFUL LOCAL AUTHORITY ACTION: BEYOND 'BEST PRACTICE'

One of the key findings of the research led by CSE in 2005 was that, while there existed examples of excellent local action on climate change and sustainable energy, these were the exception rather than the rule. More importantly, behind each of the good examples were invariably the efforts of a 'wilful individual' working within the local authority. The commitment, knowledge and doggedness of these individuals, more often than not middle-ranking officials rather than those in leadership positions, have been the key to success, rather than any broader national policy or programme. Their successes were generally as a result of them tackling obstacles and resisting objections to their actions which, at the time, sat outside the explicit remit of local authorities. The work of these 'wilful individuals' has, over the last decade, silently dominated case studies written up in Government publications on climate change, itineraries for ministerial visits and the content of the 'best practice' advice programmes which have been the performance improvement process of choice in the UK for at least the last 20 years.

However, there has also been a strong tendency to see the conditions which these characters have managed to create – strategic coherence, political and senior management support, resource prioritization and so on – as the conditions required for success. Thus, best practice advice tends to stress the importance of these conditions to the achievement of best practice. And performance improvement programmes, such as those offered by the Energy Saving Trust, typically start with an attempt to engage senior officials.

There is undoubtedly some truth in this advice. It is unusual to find effective and sustained local authority action on some aspect of carbon management without these conditions having been established. However, this may be a case of mistaking 'cause' for 'effect'. It is even more unusual to find effective and sustained local authority action on some aspect of carbon emissions reduction without a 'wilful individual' being involved. It is much more likely that these factors are not the 'pre-conditions for success' but are actually the symptoms – or results – of a successful wilful individual who has been at work for several years creating the right conditions for his or her work to succeed.

For the above reasons therefore, best practice is difficult to replicate – and as such is often disempowering for those less wilful officials to whom it is presented as a blueprint for their own work since it involves far more effort and change than they can contemplate achieving. In direct contrast to much guidance and advice for local authorities on climate change – which tends to focus primarily on securing senior management commitment and 'declarations' of intent at an early stage (for example, the Nottingham Declaration, the IDEA Beacon Council Toolkit) – this chapter suggests that a better approach would be to focus on 'a few people doing a few things effectively' as the starting point for improving performance. In this approach, it is assumed that strategic buy-in and an emergence of cross-authority working on carbon reduction follow as a *consequence* of incremental improvements rather than lead as their cause. As outlined below, this has implications for the next steps which local authorities should be taking towards improving performance on reducing emissions within their communities.

It is currently too early to assess the effect of the introduction of carbon emission reduction into the local authority performance framework in England, where it only came into force in April 2008. The new performance indicator on cutting carbon emissions at the local level was specifically designed to establish conditions within which less wilful individuals working in local government could be more effective with less effort. The priorities of senior officials and political leaders in local authorities have always tended to be shaped by the performance framework used by the Audit Commission. It is hoped that the introduction of carbon emission reduction into this framework will create the demand and the space for local authorities to start moving towards a comprehensive, 'whole-authority approach' to tackling carbon emissions – one which is patently required by the complex nature of climate change itself.

IMPROVING LOCAL AUTHORITY PERFORMANCE: FROM 'EASY WINS' TO 'LAST IMPACTS' WITH 'BIG STRIDES'

The importance of incremental improvement above 'strategizing' formed the basis of further research by CSE (2007), where it sought to identify what this would mean for local authorities at different current levels of performance. The study examined the Local and Regional Carbon Management Matrix from the perspective of local authorities which were at different stages of improving their performance.

There is a reasonable argument to suggest that local authorities which have done little to date and rank 'weak' or 'fair' on most levers in the matrix will need to look for relatively straightforward, limited resource activities which make a reasonable impact quickly. These can be characterized as 'EASY WINS'. These will mainly be found in the steps in the matrix taking an authority from 'weak' to 'fair' or 'fair' to 'good'. For example, a fresh focus on an authority's own energy management is likely to generate cost savings which can reasonably quickly support further improvement work.

Using the matrix as a guide suggests that local authorities which have understood the potential of their role and already achieved some of the 'easy wins' will need to be planning out 'BIG STRIDES' to create real impact through effective resource allocation and strategic development. These will mainly be found in the steps taking an authority from 'fair' to 'good'.

Finally, those authorities which have already made good progress need to be looking to embed better performance on local carbon management across all of their activities and allocating resources to fit this sense of priority. These will tend to be 'good' to 'excellent' steps in the matrix, leading to 'LASTING IMPACTS'. The 'Top 5' of each of these categories is detailed in Table 5.1. All of the 'Easy Wins' are actions which could potentially happen through the initiative of one or two people 'getting on with it' and drawing in expertise and resources from other agencies (for example, energy advice centres) or insulation programmes (for example, Warm Front and energy supplier CERT schemes). It is not until a relatively late stage in the performance improvement process that the full strategic approach across the local authority is expected to be embedded.

CONCLUDING REMARKS: THE NEED TO UNDERSTAND THE PROCESS OF IMPROVING LOCAL AUTHORITY ACTION ON CARBON EMISSIONS

The role of local authorities in galvanizing action to tackle climate change has now been accepted, defined and firmly embedded in the UK policy framework for delivering its ambitious carbon reduction targets. This chapter has argued that there is now a need to understand much better how to secure the necessary improvement by local authorities in their performance in this role so that good practice becomes the norm rather than the exception.

As identified by research carried out by CSE et al. (2005), there is currently a paucity of data on the effectiveness (or otherwise) of local authority action to curb carbon emissions at a local level alongside the associated costs of such action. The introduction of the new local authority performance indicator on community carbon emission reduction

Table 5.1   Top 5 'easy wins', 'big strides' and 'lasting impacts' (from CSE, 2007)

EASY WINS	BIG STRIDES	LASTING IMPACTS
Promotion of available grants and energy efficiency schemes (domestic) (Warm Front and energy supplier certificate schemes)	Active engagement (in partnership with other LAs and agencies (rural and mixed)) to establish grant-scheme to drive take-up of insulation and heating improvements	Set carbon reduction strategy and ambitious targets within Local Strategic Partnership (and Local Area Agreement) with cross-sectoral delivery plan
Engagement and promotion of local energy advice service (Energy Saving Trust advice centre)	Establish plan for achieving high thermal standards in decent homes plans	Clear and tailored strategy to improve domestic energy efficiency with regular public promotion to demonstrate commitment and political leadership
Establishing dedicated resource for own building energy management	Engage with Carbon Trust Carbon Management Programme or undertake own structured plan for monitoring and targeting energy waste	Involving all staff in energy and carbon management with clear investment, procurement and behavioural approaches to meet strong carbon reduction targets (including staff induction training)
Setting clear energy efficiency/carbon-based standards in procurement of own equipment, services and buildings	Developing advice, projects and neighbourhood-based approaches to reducing car transport *(mixed and urban)*. Establishing partnership with neighbouring authorities (on county basis?) to provide critical mass for procurement, improvement scheme development, advice and so on *(rural and mixed districts)*	Support, potentially with other authorities, integrated 'one-stop shop' for energy efficiency advice, including access to improvement programmes and wide-ranging referral networks

Table 5.1   (continued)

EASY WINS	BIG STRIDES	LASTING IMPACTS
Focus planning policy positively on renewable energy for larger developments *(rural)* and on-site targets for new developments *(mixed and urban)*	Establish energy efficiency and carbon reduction as priority for regeneration and economic development *(mainly urban)*. Proactive identification of renewable energy opportunities and strategy to realize (in partnership with neighbouring LAs?) *(rural and mixed)*	Develop area-based approaches to reducing car use, involving business and householders *(mixed and urban)*. Develop and train staff and members to understand renewable technologies, policies and create positive disposition *(rural)*

Note: LA = local authority.

provides an ideal opportunity to start gathering such data to enable the actions taken by local authorities to be compared with the real emission reductions achieved (as measured by the indicator itself). Analysis of such data would start to reveal which actions by local authorities appear to be associated with higher levels of emission reductions. Thus 'best practice' advice would start to be informed more by robust evidence rather than anecdote.

In addition, it is suggested that research into the processes which lead to genuine performance improvement in local authorities in relation to carbon emission reduction is also needed. While the analysis outlined here suggests an incremental 'bottom-up' approach would prove more effective, it is argued that it will be important to develop an evidence base which goes beyond current 'observation-based supposition' in order to reveal the *real* success factors for improving local authority action.

NOTES

1. DTI official, personal communication with author, April 2005.
2. Which have themselves evolved over many decades of defining and redefining the optimal division of responsibilities between local and national government.

REFERENCES

CSE (Centre for Sustainable Energy) (2006), *The Local and Regional Carbon Management Matrix*, Bristol: CSE, accessed at www.cse.org.uk/projects/view/1082.

CSE (2007), *Council Action to Curb Climate Change: Key Issues for Local Authorities*, report for the Local Government Climate Change Commission, London: LGA.

CSE with CDX (2007), *Mobilising Individual Behaviour Change through Community Initiatives*, report to Defra, DTI, HM Treasury, CLG and DfT, London: Department for Environment, Food and Rural Affairs, accessed at www.cse.org.uk/pdf/pub1073.pdf.

CSE with Impetus Consulting and QE2 (2005), *Local and Regional Action to Cut Carbon: An Appraisal of the Scope for Further CO2 Emission Reductions from Local and Regional Activity*, report to DEFRA for the UK Climate Change Programme Review, Bristol: CSE.

Futerra (2005), *UK Communications Strategy on Climate Change: Recommendations to Defra, DTI, Carbon Trust, Energy Saving Trust, Environment Agency and UK Climate Impacts Programme*, London: Futerra.

HM Government (2000), *UK Climate Change Programme*, London: HMSO.

HM Government (2003), *Our Energy Future: Creating a Low Carbon Economy*, London: HMSO.

HM Government (2004), *Review of UK Climate Change Programme – Consultation Paper*, London: HMSO.

HM Government (2006), *Climate Change Programme: The UK Programme 2006*, London: HMSO.

HM Government (2009), *The UK Low Carbon Transition Plan: National Strategy for Climate and Energy*, London: Department for Energy and Climate Change, accessed at www.decc.gov.uk/en/content/cms/publications/lc_trans_plan/lc_trans_plan.aspx.

6. Mobilizing sustainability: partnership working between a pro-cycling NGO and local government in London

Justin Spinney

INTRODUCTION

In the move towards a low carbon world modern forms of mobility have been increasingly problematized, particularly automobilities (Bickerstaff and Whitelegg, 1987; Walker, 1999; Vigar, 2002; Bohm et al., 2006; Bonham, 2006; Horton, 2006, 2007; Merriman, 2009). As awareness and evidence of the global and local environmental (not to mention bodily) consequences of automobility have grown, particular groups have consistently positioned the bike and cycling as a panacea for modern urban ills, framing it as environmentally benign and healthy in opposition to and as a replacement for the car. However, whilst there have been various booms in cycling as a 'leisure' practice, against a background of increasing car ownership, its popularity as a mode of everyday movement in the UK has been in gradual decline[1] since the Second World War to the point now where in Greater London (UK) less than 2 per cent of all 'utility' journeys are made by bicycle (3.7 per cent in inner London) (Transport for London (TfL), 2008, p. 1).

Part of the reason for this decline, it is argued, is that London's car-centric infrastructure militates against cycling. In order to rectify this situation successive governments have focused upon improving the infrastructure[2] for cycling in order to promote the bike as a means of transport. As a result attempts have been made in recent years to provide this infrastructure (in the form of the London Cycle Network – LCN – in London). In order to ensure that the proposed routes and improvements are what cyclists require, policy makers (such as TfL) have been encouraged to include the cycling community in the decision-making process. This has largely been attempted through partnership working with key stakeholders (such as the London Cycling Campaign – LCC), forming policy communities with responsibility for particular geographic areas and spatial practices.

This move to partnership working is part of a broader post-modern 'collaborative turn' in planning (Davidoff, [1965] 2003; Brindley et al., [1996] 2003; Healey, [1996] 2003, 1997; Khakee, 1998; Tewdwr-Jones and Allmendinger, 1998; Forester, 1999; Fainstein, 2000; Bickerstaff et al., 2002; Verma, 2007), which has sought to involve users in the design and materialization of projects. Petts and Brooks (2006, p. 1046) note that 'bringing the lay public into environmental decision-making processes is a means of ensuring that "plural voices" are heard and acknowledged'. As Hajer and Kesselring (1999) note, three of the theoretical arguments for increased public participation are that it will enhance democracy, generate new forms of knowledge and bring new institutions for more effective governance. The policy communities formed through such partnerships attempt to represent the interests of a wider community of practice; in this case that of the everyday commuter cyclist.

However, many authors have called into question the efficacy of the post-modern approach to consultation, suggesting that whilst there is much activity on the surface there is often little difference in the end products (Bickerstaff et al., 2002). Indeed, Healey ([1996] 2003) argues that often the best that can be hoped for is that new issues at least come to the attention of planners and policy makers. Lefebvre (1991) argues that the failure of advocacy planning is because the specialists who speak on behalf of specific groups of users lack the language and the right to do so. For Lefebvre, 'the silence of the users is indeed a problem – and it is the entire problem' (p. 365).

However, it is perhaps not only the silence of users which is a problem. Rather, as I go on to argue in this chapter, what is perhaps of more concern is the problem of translating knowledges within a given policy community. Habermas (1991) notes that the current public sphere is characterized by forms of strategic power where an asymmetric power distribution means that certain groups can effectively impede other groups from realizing their interests (in Skollerhorn, 1998, p. 556). What I am concerned to demonstrate in this chapter are some of the ways in which asymmetric power relations evolve premised upon particular (expert) forms of knowledge and practice. I contend here that the policy community formed in this instance between central government (TfL), local government and a pro-cycling non governmental organization (NGO) – the LCC – comes to value established 'expert' planning practices over the embodied 'lay' practices and knowledges of the LCC and those it seeks to represent.[3] As I demonstrate however, rather than be excluded from debate, in order to be heard as a legitimate voice the lay understandings of the LCC are seen to evolve into more expert knowledges and practices in line with those existing within TfL and the planning profession more generally. As a result it is argued that the trajectory[4] towards expert forms of knowledge may leave groups such as the LCC in a position where they are less representative of the wider lay group which they purport to represent. Thus the process of 'partnership' is ironically one of the things that perhaps moves interest groups further away from the community of practice that they seek to represent. Whilst the communicative turn and the communities it produces seek to give voice to alternative practices and knowledges, in this instance the process appears to exclude or translate these alternative voices according to a dominant and abstracted idea of cycling as a practice. As Booth and Richardson note, true participation requires more than just the standard tools and approaches (2001, p. 146) and this is demonstrated to be problematic in this case.

MAKING SPACE FOR CYCLING

The acceleration of automobility has brought with it a substantive reshaping of urban environments and rising levels of traffic and congestion. Unsurprisingly, widespread concern has increasingly been voiced (particularly since the 1970s) regarding the consequences of this shift (Horton, 2006, p. 43; see also Vigar, 2002; Horton, 2007). The 1990s (and in particular the New Labour 'New Deal on Transport' White Paper of 1998) saw a theoretical emphasis on reducing the harmful environmental effects of mobility. In London recent environmental (and economic) imperatives have been materialized via the Greater London Assembly and TfL as an increased emphasis on public transport, walking and cycling (Hine, 1998, p. 144), and congestion charging for cars and commercial vehicles at times of peak traffic flow in central areas. Based upon such policies efforts have been made in recent years to increase the numbers of people cycling in the capital. To date efforts in the UK have primarily been aimed at increasing access to cycling as a transport practice and have taken the form of altering infrastructure to facilitate cycling (with mixed results), and to a lesser extent giving people the skills and confidence required to cycle in the form of training and advice.

Whilst the planning process in relation to cycling varies greatly from country to country and between regions, the primary focus in terms of infrastructural change in London is the London Cycle Network Plus[5] (LCN+), around which most cycle provision centres. Originally formulated in the 1980s, it was not until the early 1990s that the LCN started to become a reality. The LCN+ is a 900 km network spanning 33 London boroughs (TfL, 2007a, p. 2), envisaged 'as a means of concentrating activity on a core strategic network' (Transport Research Laboratories (TRL), 2005, p. 16). This essentially means that particular core routes across London deemed most appropriate for cycling have been selected and funds from TfL are focused on improving the conditions on these routes for cycling.[6] By 2007, approximately half of the 900 km was deemed by TfL to be completed, with the rest due for completion by 2010 (TfL, 2007b, p. 2). The London-wide budget for the LCN+ in 2005/6 was £11.47 million ('London Cycle Network Plus', 2006).

In order to help specify and plan the LCN+ it would seem desirable to involve end-users in order to ensure that the routes and treatments meet the everyday needs of cyclists (see for example Crewe, 2001; Creighton, 2005; Fiskaa, 2005). However, despite directives to bring publics into the orbit of decision-making, for a number of reasons which have been well documented this rarely seems to happen (Brody et al., 2003; Abram and Cowell, 2004; Innes et al., 2007). Where public participation is attempted it often falls far short of its goals, as Booth and Richardson (2001) have noted in relation to the trunk roads consultation process. Here they observed that in general the process excluded publics from defining the problem, with their input limited to influencing alignment, amenities and junction arrangements (p. 143). Consequently there are very few instances of successful public and community participation in the literature[7] and certainly Brabham (2009) notes the continued inability of planners to adequately involve the public.

By way of compromise, in order to help specify and appraise plans for the LCN+ and ensure that it better meets the needs of end-users the central stakeholders (TfL and relevant local authorities) work with a number of different stakeholder groups which contribute advice and viewpoints on what is required. For example, for the formulation of LCN+ routes, principal stakeholders in communication with TfL were (and are) Living Streets, the Royal National Institute of Blind People and the LCC[8] (TRL, 2005, p. 16). For more detailed consultation necessary for the Cycle Route Inspection Meeting (CRIM) and Cycle Route Implementation and Stakeholder Plan (CRISP) processes (which I discuss in more detail later), the key stakeholder for cycling in most boroughs is the LCC[9] (see for example TRL, 2005, pp. 16–17; City of London Borough Council, 2005, p. 31; Hammersmith and Fulham Borough Council, 2005, p. 76), with additional but less consistent input from two other national advocacy groups: the National Cyclists' Organisation (formerly the Cyclists' Touring Club) and Sustrans.

PARTNERSHIP AND POLICY COMMUNITIES

The move to work in partnership with interest groups and outside stake-holders is part of a wider communicative turn in planning (Healey, [1996] 2003, 1997, 1998, 2007; Khakee, 1998; Forester, 1999; Rydin, 1999; Verma, 2007; Brabham, 2009). McDonald (2005) states that partnership working is characterized by network forms of governance which attempt to involve normally excluded actors and foster trust, equality and reciprocity, and represents a more democratic, participatory way of dealing with 'messy' issues such as the environment. Healey ([1996] 2003) appears to suggest (following Habermas' original formation) that with sufficient goodwill there can be equitable representation of all within the public sphere. In relation to transport planning, Healey (1997) argues that the 'task is to bring together stakeholders in a variety of arenas and to manage the discourse so as to identify commonalities and overcome conflicts and barriers to action' (in Rydin, 1999, p. 475).

The main stakeholders in these policy communities are increasingly NGOs (Vigar, 2002). In the context of the political transformations of the 1970s, where the state gradually ceased to be the supreme regulatory body, environmental NGOs have gained increasing prominence as part of the new multi-layered and multi-actor policy networks and have been instrumental in bringing environmental and social issues to the attention of the public (Arts, 2002, p. 29). Until recently, the relationship of NGOs with other actors – particularly business – had been characterized as antagonistic (2002, p. 26). As Arts notes, organizations such as Greenpeace have in the past been more confrontational, but are increasingly in dialogue and partnership with business and government (2002, p. 27). Certainly Skollerhorn notes that NGOs increasingly operate as bureaucratically as the institutions they oppose, and consequently the use of NGOs by government and business has been a key element of privatization in the UK (1998, p. 557). Increasingly NGOs have occupied a more pragmatic policy space forming partnerships with government and becoming proxies for wider communities.

This new pluralism has undoubtedly resulted in story lines outside of statist and corporatist interests being heard more and more. Skollerhorn observes that NGOs have been essential to democratic debate and learning because they enable it to occur free of the influence of the institutions of government and business. However, as Booth and Richardson remark, 'stakeholder groups who become consultees in decision-making cannot be wholly representative: they will always provide a partial view' (2001, p. 147). They go on to note that stakeholder groups have become a convenient way of packaging and representing a less than homogeneous public interest (Booth and Richardson, 2001, p. 148).

Whilst the inclusion of NGOs may in principle lead to a more inclusive consultation process, this requires an 'ideal' networked public sphere free from strategic power and hierarchical relations, and there are a number of reasons why this may not occur. Whilst such partnerships can be viewed as emancipatory through the empowerment of traditionally excluded social groups, they may also represent an erosion of democratic decision-making through the elevation of non-elected representatives to speak on behalf of particular groups (Lowndes and Skelcher, 1998, p. 316). Even if we accept that the NGO stakeholders that form part of a policy community are broadly representative of the wider community, as Lowndes and Skelcher note, the nature of the partnerships formed by no means guarantees that relations between actors are conducted on the basis of mutual trust and reciprocity (1998, p. 314). They go on to observe how networks which initially characterize many such relationships are often seen as exclusionary because new (and particularly smaller) actors cannot break into them (p. 322). Moreover, Lowndes and Skelcher showed that perceptions and stereotypes about particular organizations prevented relationships and trust from being established, and that hierarchies often formed within networks based on how well resourced particular organizations were, all of which served to further exclude particular actors (usually voluntary and community organizations) (p. 323).

McDonald says that far from their ideals of trust and equality, the public spheres created through such partnerships are more often characterized by bargaining, instrumentalism and pragmatic compliance (2005, p. 581). Healey goes further, suggesting that partnerships 'become merely channels for the reconstitution of local corporatist elites, colonising the institutions of government for their benefit' (1997, p. 237). Confirming this, Arts states that power relations are often uneven and this can lead to unstable coalitions (2002, p. 35).

Of particular relevance to the ways in which power relations are enacted is the division between lay and expert knowledge and practice. Expert knowledge is perceived to be derived from verifiable empirical observation and distinctive techniques whilst lay knowledges are perceived to be based in everyday, casual common-sense understandings (Petts and Brooks, 2006, p. 1046). Booth and Richardson note the entrenched tendency for planners and engineers in local authorities to position themselves as the experts with best knowledge of the problems and solutions (Booth and Richardson, 2001, p. 148). Hou and Kinoshita (2007) and Innes et al. (2007), for example, found that the formal nature of planning meetings could work against citizens contributing and seeing themselves as part of the solution (in Brabham, 2009, p. 245). However, whilst many accounts highlight the failure of public participation and partnership working due to power imbalance, very few actually detail the ways in which power is enacted in and through divisions in knowledge.

What I am concerned to show in this chapter then are some of the ways in which power is enacted and partnerships evolve.[10] In doing so I attempt to demonstrate the dynamic nature of actors and policy communities as a result of partnership working. In line with Dudley (2003) I argue against seeing policy communities as stable, suggesting instead that they are constantly evolving due to the input of different stakeholders and broader contexts.

TRANSLATING PRACTICES

The planning process in relation to cycling has historically been dominated by expert knowledge and particular practices such as drawing and modelling (Gehl, 2001, p. 41; see also Hill, 2003; Imrie, 2003). These practices have generally been of a meta-scale and technical in nature; large data sets and models used to analyse, predict and provide (Bannister, 2002, p. 17). This point was reinforced in my interviews with transport professionals, who cited the key tools used to understand the movements of cyclists as traffic counts and observations, computer modelling, Computer Aided Design and 'best practice' guidelines, particularly the London Cycle Design Standards. Just as the rationale for planning has remained largely consistent, so too, it would seem, have the tools. Certainly it would seem that with its emphasis on the calculable, modelable and predictable, transport planning is still dominated by forms of expert knowledge and practice:

> We use a number of things to research what's needed. For example Vehicle Swept Paths (VSP) so that we can see the path of the vehicle so you can choose various types of vehicle and put them through a junction on a plan and see what clearance spaces it needs to get through. (Engineer 3, 01/07/05)

> Well those [techniques] are common right the way across engineering. If you take a drawing in engineering they draw the road but none of the behaviour of road users. So the word assumption is quite appropriate but also there's a lot of direction of people to behave in a certain way. If you can't predict then you try and direct. (Planner 1, 24/06/05)

As these comments attest, modelling and drawing form the basis of expert knowledges through which material solutions for cycling are constructed. In these instances modelling is used as a way of knowing how people move and proposing appropriate solutions. Such tools however are not simply practical; they also serve to position the planning profession as experts. As Burgess et al. note, experts 'gain authority from relying on scientific models and experiments to define appropriate management practices' (Burgess et al., 2000, p. 120). However, whilst transport professionals see these as legitimate ways of knowing and managing cycling, the use of such tools and the resulting constructions of cycling were problematized by many LCC campaigners:

> What they [planners and engineers] do is they're used to drawing plans on maps and not looking at how cyclists and other people interact with the space. The planners are all people who can use Autocad or whatever and do traffic counts but they don't just watch what happens or cycle on it themselves. (Campaigner 1, 16/02/05)

> Road design is an abstract thing and when you look at diagrams of junctions and stuff and flows and desire lines it's all very pure and beautiful but in no way can it or does it represent the randomness or the total need to be aware and the general chaos, and it is chaos; the reversing lorries, the u-turning taxis, the muppet cyclist with the shopping wrapped round the front wheel with headphones on. (Campaigner 2, 09/08/05)

These examples contrast the expert and abstract knowledges of transport professionals with the lay knowledges of everyday cyclists, suggesting that expert knowledges cannot adequately comprehend what everyday cycling is like or propose appropriate solutions. These criticisms are related to the idea that cycling is a multi-sensory and moving experience, a situation that the tidy and relatively static world of Autocad and computer modelling fails to recognize. As Cresswell (2006) notes, mechanistic constructions aim to make difference irrelevant, suggesting instead ideal movements which fail to consider 'the different ways in which people are mobile or immobile' (p. 29). Similarly Imrie (2000) states that current 'wisdom' seeks to accommodate predictable and productive mobility rather than conceiving of mobility as a messy, unpredictable and dynamic reality (p. 1644). As a result, there is a tension between how the cyclist is discursively framed through expert knowledge and tools, and how they are framed through the everyday practice of cycling.

It is here that the importance of the CRISP and CRIM in the consultation process requires some contextualization. Once a route has been designated for 'improvement', a more detailed process begins based upon the Department for Environment, Transport and the Regions (DETR) Cycle Audit and Review Guidelines (Hammersmith and Fulham Borough Council, 2005, p. 74). The role of the CRISP is to provide a framework to assess the quality of provision, propose remedial works, specify the consultation process and provide standard documents on which to base proposals (TRL, 2005, p. 17). This is followed by the CRIM, which is organized once a route has been identified and entails the proposed route being ridden by stakeholders ('London Cycle Network Plus', 2006). According to TfL the stated aim of the CRISP/CRIM process is:

> to create an environment for people with cycling expertise, local knowledge, and often widely diverging perspectives, to work together to identify common issues, concerns and priorities for the strategic cycling corridors (LCN+), and specifically to engage in a practical process which would result in agreement on preferred route alignment and link treatment and next steps. (TfL, 2007b, p. 4)

As this suggests, many of the shortcomings of the Transport Planning Model are in part recognized by transport professionals (O'Flaherty, 1997), and in the contemporary context of London these modelling and representational strategies are also tempered with the input of specific stakeholder groups through the CRISP and CRIM processes. The goal here is that the quality of planning outcomes will be enhanced through the inclusion of new ideas and knowledge (Booth and Richardson, 2001) as a result of the CRISP and CRIM.

The CRISP and CRIM initiatives are an attempt, as Booth and Richardson note, to involve transport users in order to gain 'a better understanding of the fine grain of everyday life to develop a richer understanding of travel needs, patterns and behaviour' (Booth and Richardson, 2001, p. 143). As a result the CRISP and CRIM should in principle facilitate the inclusion of everyday cycling knowledges and practices. However, in the eyes of numerous LCC campaigners the CRISP and CRIM apparently fall short of what is expected because they continue to marginalize the knowledges and practices of cyclists in a number of ways. The first way in which this is done relates to the lack of consultation in defining a given route in the CRISP process, as these LCC campaigners noted:

> The process that we were involved in presumably corresponded to the description in the CRISP document. The CRISP document assumes that, at the start of the process, an LCN+ link plan exists, with the route having been fixed. The route may have been fixed, but it was blindingly obvious that the route choice in our example was so poor as to be unacceptable, even though it had, presumably, been found acceptable by the borough, found acceptable by the sector committee, found acceptable by TfL, and found acceptable by at least one consultant, and perhaps more. (Campaigner 3, 28/05/03)

> It has come to my attention that the borough is intending to cut back the bus lanes all along its side of the A5 from Edgware to Cricklewood. They are essentially going to remove all stretches of bus lane near junctions to 'improve traffic flow'. This information came in a proposal for a traffic order sent to

us, which states 'Any person wishing to object to the proposed order should
send a written statement explaining their objection to the Design Team at
the above address within 21 days of the date of this Notice'. So this is not a
consultation and there seems to have been no consultation. (Campaigner 4,
27/02/09)

In these examples (the second of which relates to a junction treatment
where the borough rejected the use of the CRISP process outright for its
own policy reasons – a key problem in itself), much of the work of route
selection and design appears to have been decided already and the role of
the LCC is limited to commenting on a route that has been pre-determined
by a number of 'expert' groups: the borough, sector committee, TfL and
the consultant. Expert knowledge pre-defines what the route will be and lay
input is sought only at a later date. This strongly suggests that the forms of
knowledge and points of view that the LCC will provide as a stakeholder
are at the very least under-valued by the dominant stakeholders. When the
LCC does get a say in the planning process, two more examples raise the
question of how seriously its input is taken by the boroughs and TfL:

> I felt that at both meetings the organizers wanted just 'good news' about cycling
> and that anything that was other than this or questioned the status quo/wisdom
> of TfL was not really acceptable. (Campaigner 5, 08/05/07)

> At the last CRISP we attended we wrote our own report, and asked for it to be
> included as an annex to the real report. It's there. I doubt if it will have much
> influence, but it does make clear, if you read it, that what we think does not bear
> much relationship to what the consultant thought. (Campaigner 3, 07/05/07)

As is evident in these examples there is the belief amongst many LCC
campaigners that their lay views are not valued in the way that the expert
knowledge and 'wisdom' of consultants and TfL are. Indeed, in the second
example the views of the LCC are only an annex in the main report. Thus
even when the views of cyclists are sought, they are given marginal status
as less important than the official and unquestionable understandings
produced by the experts. This echoes the work of Phillimore and Moffatt
(2004), who have demonstrated the tendency of government to resist local
framings of the environment and contexts which disagree with its own.
A final set of accounts relates to the exclusion of embodied practices as a
form of knowledge:

> When there's a grassed area and they're going to put paths, what they do is they
> let people walk the area to see where it gets worn and then put the path down
> and what they should do is realize where cyclists are going to go do the same.
> (Campaigner 1, 16/02/05)

> There seemed to be considerable reluctance among TfL and the consultants to have actual cyclists along. There was an obvious wish [at the CRISP] to avoid facing the fact that different cyclists, or even the same cyclist at different times, might have different opinions about their optimum route. [. . .] The bureaucrats, with their request that in future only one cyclist appear, are in effect trying to ensure that they never hear the diversity. (Campaigner 3, 28/05/03)

The first comment highlights the lack of embodied knowledge in the process, suggesting that routes are determined not by what people do but what planners would like them to do. The second comment demonstrates that whilst CRISP consultation is in theory open to all, it is evident that only a limited number of stakeholders are actually welcome. By asking that only one cyclist be present to ride the route, any diversity of opinion and knowledge is written out. As a result, what is excluded from debate are the diverse embodied knowledges gained by cyclists in the form of everyday practices and experiences of cycling particular routes on a regular basis. Whilst the CRISP and CRIM purport to be a platform to facilitate the inclusion of these forms of everyday knowledge, the reality appears somewhat different. As Booth and Richardson note, 'there are real difficulties here in developing viable methods which allow a wide range of interests, including lay people, to engage in debates or even participate in decisions about "big" transport issues' (2001, p. 144). Despite attempts to the contrary, everyday cycling practices are still 'left out in the cold if their message is "inappropriate"' (p. 147).

BECOMING EXPERT

As a result of the exclusion of their knowledges and practices, in order to be taken seriously campaigners have increasingly adopted the practices and language of transport professionals. For example the LCC has a dedicated Cycle Planning and Engineering Group which shares and develops knowledge in this field. In addition, many of its members do or have worked in the field of transport planning and engineering for borough councils or contractors (Campaigner 7, 21/05/08). It is clearly the case that members of the LCC are familiar with the practices of planning, something particularly evident in the language that they employ and the detailed knowledge they display of policy documents and processes:

> In our area this route has been redeveloped and is unrecognisable from before. And when was the CRIM/CRISP? I thought the first CRIM/CRISP was the A13 [06] one back in May '03. Who was on the route 55 CRIM? Has anyone got any record of it formal or otherwise? [. . .] Why is the CRIM/CRISP for . . . Cycle

route (55) not going any further than the borough boundary near the bottom of Uphall Rd (LCG 6 37Rc)? This will leave a mile of the route unCRISPed and the route only makes sense if it goes to the Station. (Campaigner 6, 30/03/07)

One campaigner went on to explicitly acknowledge the importance of using the tools of planning in order to be heard:

Of course it could be that what the model claims will happen is unlikely to reflect reality. That makes it your word against the 'expert's'. So it's best to have some data, or research, or a model of your own, or a counter example (preferably with lots of glossy colour photos of it in action), or something, to back up your case. (Campaigner 3, 08/09/07)

Petts and Brooks (2006) talk of a deficit model in planning whereby publics and stakeholders lack expert knowledge and must be informed in order to be able to take part in debate. However, these accounts illustrate that it is not so much the lack of knowledge but the form that it takes that excludes the LCC from the debate. As a result it would appear that campaigners become professionalized and 'expert' because they are forced to learn the dominant language of planning to be heard as a legitimate voice in the debate:

Campaigners are members of the public and . . . have (I hope) built up a certain level of expertise, such as the 'person in the street' hasn't got, which should inform their contribution. [. . .] If the council is basically saying that it only wants to consult people who are inexpert/inexperienced, 'Campaigner' becomes a disparaging term, rather like 'expert' can be ('so-called "experts"'). [. . .] In my campaigning, therefore, I can only ask for what I think cyclists want, rather than what they tell me they want. All the more reason to build up as much knowledge and expertise as possible. (Campaigner 6, 23/11/01)

There is an increasing amount of consultations, meetings, conferences. Being reactive to these should not automatically be seen as the best way to campaign. The LCC seems increasingly stuck in a groove that the way to do things is to go to meetings and talk to important people [. . .] I for one feel I have been far too bogged down in CRISPs in the last two years for very limited results and with hindsight wish that I'd campaigned to improve training, knowledge etc of engineers and consultants instead. (Campaigner 5, 08/05/07)

These two accounts illustrate different takes on the desirability of expert knowledge. The first account suggests that it is important for campaigners to become experts in order to adequately represent what they think cyclists want and be taken seriously by borough officials. Goven (2003) and Irwin (2001) note that such processes often promote institutional capture and reactive citizenship rather than creative citizenship. The second account suggests that some campaigners are uncomfortable with the fact that they have had to become experts in planning practice, seeing it as a diversion from the real business of campaigning. However, despite the tensions it is clear from these accounts – and particularly when considering the origins of the LCC as a group employing 'guerrilla' tactics to get its message across[11] – that inclusion in this particular policy community with TfL and the borough councils has changed the approach and practices of the LCC to be more in line with those of the dominant groups. The emphasis which these campaigners put on being experts echoes the point made by Burgess et al. that in order to 'achieve success, other actors' worlds must be colonized. Actors become powerful through their abilities to enroll others in a network and to extend their network over greater distances' (Burgess et al., 2000, p. 123).

In these examples, the LCC in order to be able to 'campaign' is enrolled into the dominant world of established planning and in the process is transformed as an institution. Jamison (1996) for one advocates a processual view of the development of environmental NGOs which is backed up by the evidence here. He argues that the concept of agreement in such debates needs to be understood not just as a convergence of interpretations, but rather as the result of the transformation of social movements and institutions (p. 238).

CONCLUSIONS

Whilst partnership working has the potential to bring more voices and storylines into the planning process and policy communities, the evidence here suggests that the relationship between actors remains hierarchical and falls short of a communicative rationality (Healey, 1997, 1998). As McDonald (2005) notes, the current public sphere is one still dominated by what Habermas termed an instrumental rationality where actions are dictated by goals set by the state and the market of material reproduction and maximizing production (p. 584). Burgess et al. give numerous examples where despite more innovative participatory forums (of which I would suggest the CRIM and CRISP are examples), an instrumental and scientific rationality remains dominant (1998, p. 1448) which constructs the public in particular ways (p. 1449).

Despite the introduction of the CRISP and CRIM processes it is suggested that the everyday user and lay knowledges are still marginalized within planning discourse. Whilst there are of course issues of governance at work here in the failure of the CRISP/CRIM process,[12] this chapter has tried to illustrate some of the other ways in which the process fails;

primarily by excluding stakeholders from meaningful engagement, marginalizing their voices in documentation, or by rejecting the legitimacy of their lay knowledges. As Vigar (2002) notes, the transport policy communities thus formed are often of limited membership either because of deliberate exclusion on behalf of the policy community or through an 'unconscious conspiracy' where views are not actively sought from certain stakeholder groups (p. 209). The absence of particular storylines and the subsequent dominance of others are problematic and serve to reinforce a narrow idea of possible solutions. As a result it is questionable, as Kerr et al. (2007) have noted, how successful processes which seek to bridge the expert–lay divide actually are.

However, such a reading risks suggesting that groups such as the LCC are excluded from debate altogether. I would argue that this is not entirely the case, largely because stakeholders such as the LCC undergo a process of transformation through taking on board the more expert knowledges and practices of the dominant stakeholders in the policy community. Wynne (1996) and Jamison (1996) both critique environmental NGOs for their increasing reliance on scientific discourses and expert knowledges. However, it is perhaps not so much that this is what these organizations desire (though they may well), but rather that in the face of such uneven power geometries, rather than be excluded from debate, becoming conversant with expert forms of knowledge becomes the only way in which such groups can maintain legitimacy in the policy community. However, as Booth and Richardson (2001) note, as a result, 'inclusivity is not being created: rather, exclusivity is being redefined' (p. 148). Certainly, as Brabham (2009) notes, 'the very presence of special interest groups in the planning process, who show up to planning meetings representing some facet of the public, may intimidate the average citizen with charts, maps, empirical evidence, and expert advice, thus deterring future involvement by non-experts in the community' (p. 245). This process supports Habermas' contention that NGOs tend to be transformed into institutions.

One possible outcome of this process is that NGOs such as the LCC become more expert than lay in the practices and knowledges that they use to achieve their goals. The result may be that they are not taken seriously either by the policy community with which they engage, or the wider community that they seek to represent, instead occupying a liminal policy space.

ACKNOWLEDGEMENTS

First I'd like to thank all the LCC campaigners, borough engineers and planners who gave up their time and knowledge for interview and allowed their comments to be reproduced here. A number of these campaigners deserve special thanks for detailed and insightful comments on a previous draft of this chapter. I'd also like to thank Michael Peters for his useful comments on an early draft, Phil Crang for supervising the PhD from which this chapter is drawn and the ESRC for funding that PhD. As always, any omissions and incorrect interpretations that remain are my own.

NOTES

1. It should be noted that while most nations have seen a decline in cycling over time the pattern has varied greatly in space. For example, in many European cities and states the bicycle has been given far greater prominence in transport planning and policy than in the UK, with the result that cyclists are far less marginalized in debates and everyday practice.
2. Infrastructure is just one way in which the landscape for cycling can be improved and it is still a matter of intense debate as to whether this is the best way to improve the predicament of cyclists.
3. It should be noted that whilst I deal here with the aggregate experiences of a number of boroughs in order to make a point, individual borough campaign groups have experienced differing levels of success and had to accommodate differing levels of expert knowledge in order to be heard.
4. For the purposes of clarity the narrative I describe in this chapter is somewhat linear. However, the reality is of course somewhat messier and more dynamic, as one campaigner noted, with engineers and planners leaving, campaigners becoming cycle officers, and policies (and mayors) changing all the time (Campaigner 7, 06/10/09).
5. The LCN+ replaced the more ambitious 3500 km LCN in 2001 with a view to producing a smaller but higher quality network ('London Cycle Network Plus', 2006).
6. If a borough wishes to improve a non-LCN+ road for cycling, provide cycling parking or implement a non-utility facility such as a BMX/Skate park, it must do so out of its own budget if it is within the council's remit, or in the case of roads such as the A4 and A40, for example in Hammersmith and Fulham, it must negotiate with TfL, which administers the 'main road' network (Hammersmith and Fulham Borough Council, 2005, p. 73).
7. Notable exceptions include the NGO Transport 2000, which in 1988 partnered with 15 authorities to deliver area-wide traffic-calming in conjunction with local communities (Booth and Richardson, 2001, p. 145).
8. The LCC was launched in 1978 (LCC magazine, December 1978, p. 1) and sprang out of a number of locally based groups which were actively campaigning for better conditions for cycling in the capital (LCC magazine, November 1988, p. 16). At this time, Friends of the Earth ran a national cycling campaign but a need was felt for a London-based organization (ibid). The origins of the LCC are closely linked to those of the green movement, particularly Friends of the Earth whose Chelsea-based publication at the time, *On your Bike*, was to be published by the LCC the following year (ibid). As Horton notes, it was during this period that environmentalist concerns, 'shifted away from the protection of particular sites and species and towards more explicit critique of specific environmentally damaging practices, such as use of the car' (Horton, 2006, p. 43).
9. Only one borough pointed out any sustained contact with a non-campaigning-based bicycle user group (Cycling Officer 4, 02/06/06). One campaigner also pointed out the lack of legal requirement in the CRISP/CRIM process to bring together all relevant

stakeholders such as the Royal Parks Authority and British Waterways (Campaigner 4, 07/10/09).
10.	The resulting narrative is based upon a series of qualitative interviews with LCC campaigners, and engineers, planners and cycling officers from 11 London boroughs conducted between 2004 and 2005. These interviews have been supplemented with more recent material drawn from the online discussion forum of the LCC. In order to protect identities, all names and boroughs have been made anonymous and places in some quotes have been removed. In line with Robinson (2001) and Sixsmith and Murray (2001) written consent has been obtained from those whose comments have been drawn from the LCC discussion forum.
11.	In its early days, the campaign was much more guerrilla minded, as evidenced by its organization of protest rides and impromptu taping of road markings and signage to streets (LCC magazine, May 1979, p. 1). Indeed it wryly noted that, 'many people have an image of the campaign as a bunch of mad radicals who cause trouble at the drop of a hat (or road plan)' (LCC magazine, April 1983, p. 1). In the past this radical tendency has been illustrated in numerous articles, many of which have lamented the sorry and ambiguous state of the law relating to cyclists (see for example May 1979, p. 3; April 1983, p. 4; February 1984, p. 2; May–June 1990, p. 15; March–April 1992, p. 9; June–July 1993, p. 13; June–July 2003, p. 8). However, this radical stance has been increasingly replaced with an arguably more strategic emphasis geared towards improving conditions within the existing legislative and operative framework. Indeed, the campaign has consciously become more mainstream over time, with one article noting how 'the LCC intends to broaden its appeal with redesign of image and the magazine – to project the image of a "modern active campaign"' (LCC magazine, Jan.–Feb. 1987, p. 12).
12.	As Campaigner 4 pointed out to me when commenting on a draft of this chapter, the lack of a legal appeals process in relation to the CRISP/CRIM was also a reason for the disenfranchisement of the stakeholder community when route and alignment decisions inexplicably went against what they had said (07/10/09).

REFERENCES

Abram, S. and R. Cowell (2004), 'Learning policy: the contextual curtain and conceptual barriers', *European Planning Studies*, **12** (2), 209–28.
Arts, B. (2002), 'Green alliances of business and NGOs. New styles of self-regulation or dead-end roads?', *Corporate Social Responsibility and Environmental Management*, **9**, 26–36.
Bannister, D. (2002), *Transport Planning*, London: Spon Press.
Bickerstaff, K. and G. Walker (1999), 'Clearing the smog: public responses to air quality information', *Local Environment*, **4**, 279–94.
Bickerstaff, K., R. Tolley and G. Walker (2002), 'Transport planning and participation: the rhetoric and realities of public involvement', *Journal of Transport Geography*, **10**, 61–73.
Bohm, S., C. Jones, C. Land and M. Paterson (2006), 'Introduction: impossibilities of automobility', *Sociological Review*, **54** (s1), 2–16.
Bonham, J. (2006), 'Transport: disciplining the body that travels', *Sociological Review*, **54** (s1), 57–74.
Booth, C. and T. Richardson (2001), 'Placing the public in integrated transport planning', *Transport Policy*, **8**, 141–9.
Brabham, D. (2009), 'Crowdsourcing the public participation process for planning projects', *Planning Theory*, **8** (3), 242–62.

Brindley, T., Y. Rydin and G. Stoker ([1996] 2003), 'Popular planning: Coin Street London', in S. Campbell and S. Fainstein (eds), *Readings in Planning Theory*, Malden, MA: Blackwell, pp. 296–317.

Brody, S.D., D.R. Godschalk and R.J. Burby (2003), 'Mandating citizen participation in plan making: six strategic planning choices', *Journal of the American Plannning Association*, **69** (3), 245–64.

Burgess, J., J. Clark and C. Harrison (2000), 'Knowledges in action: an actor network analysis of a wetland agri-environmental scheme', *Ecological Economics*, **35**, 119–32.

Burgess, J., C. Harrison and P. Filius (1998), "Environmental communication and the cultural politics of environment citizenship', *Environment and Planning A*, **30**, 1445–60.

City of London Borough Council (2005), *City of London Cycling Plan*, London: Department of Planning and Transportation.

Creighton, J.L. (2005), *The Public Participation Handbook: Making Better Decisions through Citizen Involvement*, San Francisco, CA: Jossey-Bass.

Cresswell, T. (2006), 'The right to mobility: the production of mobility in the courtroom', *Antipode*, **38** (4), 735–54.

Crewe, K. (2001), 'The quality of participatory design: the effects of citizen input on the design of the Boston Southwest corridor', *Journal of the American Planning Association*, **67** (4), 437–55.

Davidoff, P. ([1965] 2003), 'Advocacy and pluralism in planning', in S. Campbell and S. Fainstein (eds), *Readings in Planning Theory*, Malden, MA: Blackwell, pp. 210–23.

Department of the Environment, Transport and the Regions (1998), *A New Deal for Transport: Better for Everyone*, London: The Stationery Office.

Dudley, G. (2003), 'Ideas, bargaining and flexible policy communities: policy change and the case of the Oxford transport strategy', *Public Administration*, **81** (3), 433–58.

Fainstein, S. (2000), 'New directions in planning theory', *Urban Affairs Review*, **35** (4), 451–78.

Fiskaa, H. (2005), 'Past and future for public participation in Norwegian physical planning', *European Planning Studies*, **13** (1), 157–74.

Forester, J. (1999), *The Deliberative Practitioner*, Cambridge, MA: MIT Press.

Gehl, J. (2001), *Life between Buildings*, Copenhagen: Arkitektens Forlag, Danish Architectural Press.

Goven, J. (2003), 'Deploying the Consensus Conference in New Zealand: democracy and deproblematization', *Public Understanding of Science*, **12**, 423–40.

Habermas, J. (1991), 'A reply', in A. Honneth and H. Joas (eds), *Communicative Action: Essays on Juergen Habermas's The Theory of Communicative Action*, Cambridge: Polity Press.

Hajer, M. and S. Kesselring (1999), 'Democracy in the risk society? Learning from the new politics of mobility in Munich', *Environmental Politics*, **8** (3), 1–23.

Hammersmith and Fulham Borough Council (2005), 'Local implementation plan for transport 2005–2009', consultation draft, February.

Healey, P. ([1996] 2003), 'The communicative turn in planning theory and its implications for spatial strategy formation', in S. Campbell and S. Fainstein (eds), *Readings in Planning Theory*, Malden, MA: Blackwell, pp. 237–55.

Healey, P. (1997), *Collaborative Planning: Shaping Places in Fragmented Societies*, London: Macmillan.

Healey, P. (1998), 'Building institutional capacity through collaborative approaches to urban planning', *Environment and Planning D*, **30**, 1531–46.

Healey, P. (2007), 'The new institutionalism and the transformative goals of planning', in N. Verma (ed.), *Institutions and Planning*, Oxford: Elsevier, pp. 61–89.

Hill, J. (2003), *Actions of Architecture: Architects and Creative Users*, London: Routledge.

Hine, J. (1998), 'Roads, regulation and road user behaviour', *Journal of Transport Geography*, **6** (2), 143–58.

Horton, D. (2006), 'Environmentalism and the bicycle', *Environmental Politics*, **15** (1), 41–58.

Horton, D. (2007), 'Fear of cycling', in D. Horton, P. Rosen and P. Cox (eds), *Cycling and Society*, Aldershot: Ashgate, pp. 133–52.

Hou, J. and I. Kinoshita (2007), 'Bridging community differences through informal processes: re-examining participatory planning in Seattle and Matsudo', *Journal of Planning Education and Research*, **26** (3), 301–14.

Imrie, R. (2000), 'Disability and discourses of mobility and movement', *Environment and Planning A*, **32**, 1641–56.

Imrie, R. (2003), 'Architects' conceptions of the human body', *Environment and Planning D: Society and Space*, **21**, 47–65.

Innes, J.E., S. Connick and D. Booher (2007), 'Informality as a planning strategy: collaborative water management in the CALFED Bay-Delta Program', *Journal of the American Planning Association*, **73** (2), 195–210.

Irwin, A. (2001), 'Constructing the scientific citizen: science and democracy in the biosciences', *Public Understanding of Science*, **10** (1), 1–18.

Jamison, A. (1996), 'The shaping of the global environmental agenda: the role of non-governmental organisations', in S. Lash, B. Szerszynski and B. Wynne (eds), *Risk, Environment and Modernity: Towards a New Ecology*, London: Sage, pp. 224–45.

Kerr, A., S. Cunningham-Burley and R. Tutton (2007), 'Shifting subject positions: expert and lay people in public dialogue', *Social Studies of Science*, **37** (3), 385–411.

Khakee, A. (1998), 'The communicative turn in planning and evaluation', in N. Lichfield, A. Barbanente, D. Bori, A. Khakee and A. Prat (eds), *Evaluation in Planning: Facing the Challenges of Complexity*, Berlin: Springer, pp. 97–112.

LCC Magazine (various issues), London: London Cycling Campaign.

Lefebvre, H. (1991), *The Production of Space*, Oxford: Blackwell.

'London Cycle Network Plus' (2006), *London Cycling Campaign*, accessed 8 January 2007 at www.lcc.org.uk/index.asp?PageID=222.

Lowndes, V. and C. Skelcher (1998), 'The dynamics of multi-organisational partnerships: an analysis of changing modes of governance', *Public Administration*, **76**, 313–33.

McDonald, I. (2005), 'Theorising partnerships: governance, communicative action and sport policy', *Journal of Social Policy*, **34** (4), 579–600.

Merriman, P. (2009), 'Automobilities and the geographies of the car', *Geography Compass*, **3** (2), 586–99.

O'Flaherty, C. (ed.) (1997), *Transport Planning and Engineering*, London: Arnold.

Petts, J. and C. Brooks (2006), 'Expert conceptualisations of the role of lay knowledge in environmental decisionmaking: challenges for deliberative democracy', *Environment and Planning A*, **38**, 1045–59.

Phillimore, P. and S. Moffatt (2004), 'If we have wrong perceptions of our area, we cannot be surprised if others do as well. Representing risk in Teesside's environmental politics', *Journal of Risk Research*, **7** (2), 171–84.

Robinson, K.M. (2001), 'Unsolicited narratives from the internet: a rich source of data', *Qualitative Health Research*, **11** (5), 706–14.

Rydin, Y. (1999), 'Can we talk ourselves into sustainability? The role of discourse in the environmental policy process', *Environmental Values*, **8**, 467–84.

Sixsmith, J. and C. Murray (2001), 'Ethical issues in the documentary data analysis of internet posts and archives', *Qualitative Health Research*, **11** (3), 423–32.

Skollerhorn, E. (1998), 'Habermas and nature: the theory of communicative action for studying environmental policy', *Journal of Environmental Planning and Management*, **41** (5), 555–73.

Tewdwr-Jones, M. and P. Allmendinger (1998), 'Deconstructing communicative rationality: a critique of Habermasian collaborative planning', *Environment and Planning A*, **30**, 1975–89.

TfL (Transport for London) (2007a), *LCN+ High Risk Barriers Infrastructure Report*, London: Camden Consultancy Service.

TfL (2007b), *TfL Response to a Review of a Sample of Final CRISP Reports for Compliance with the CRISP Brief*, London: TfL.

TfL (2008), *Cycling in London* (October edn), accessed at www.tfl.gov.uk/assets/downloads/businessandpartners/cycling-in-london-final-october-2008.pdf.

TRL (Transport Research Laboratories) (2005), *Review of Procedures Associated with the Development and Delivery of Measures Designed to Improve Safety and Convenience for Cyclists*, Crowthorne: TRL.

Verma, N. (2007), 'Institutions and planning: an analogical enquiry', in N. Verma (ed.), *Institutions and Planning*, Oxford: Elsevier, pp. 1–16.

Vigar, G. (2002), *The Politics of Mobility: Transport, the Environment and Public Policy*, London: Spon Press.

Whitelegg, J. (1987), 'A geography of road traffic accidents', *Transactions of the Institute of British Geographers*, **12**, 161–76.

Wynne, B.E. (1996), 'May the sheep safely graze? A reflexive view of the expert–lay knowledge divide', in S. Lash, B. Szerszynski and B.E. Wynne (eds), *Risk, Environment and Modernity: Towards a New Ecology*, London: Sage, pp. 44–83.

7. Low carbon communities and the currencies of change

Gill Seyfang

INTRODUCTION

The challenge of achieving low carbon communities cannot be underestimated. While government policies set ambitious targets for carbon reduction over the next 40 years, there remains an urgent need for tools and initiatives to deliver these reductions through behaviour change among individuals, households and communities. This chapter sets out a 'New Economics' agenda for sustainable consumption which addresses the need for low carbon communities. It then applies these criteria in a critical examination of complementary currencies in the UK. These are alternative mechanisms for exchanging goods and services within a community which do not use money, and which aim instead to build local economic resilience and social capital. There have been three 'waves' of such currencies in recent years in the UK, and the chapter examines the two most recent of these, namely time banks and local money systems. The potential of these initiatives as carbon reduction tools has not previously been considered, and so this chapter offers a fresh perspective on carbon reduction, consumption and complementary currencies.

THE NEW ECONOMICS OF SUSTAINABLE CONSUMPTION

As climate change has become the most pressing environmental issue facing humanity (IPCC, 2007), the inequity of the consumption patterns which contribute to it have been thrown into relief. The risks and benefits of emitting carbon dioxide into the atmosphere are sharply divided among the world's economies, with the developed world contributing the lion's share of emissions while developing countries face the most dangerous impacts. Carbon dioxide emissions, a by-product from burning fossil fuels, are directly related to consumption levels through the energy used to manufacture, grow, transport, use and dispose of products. The UK government's Climate Change Act commits the UK to reducing its greenhouse gas emissions by 80 per cent of their 1990 levels by 2050 (HMG, 2008). However, the greenhouse gases embedded in what we as a nation consume are far greater than that in what we produce: developed countries export their carbon emissions to developing countries where manufacturing and processing occur (Druckman et al., 2008).

The Carbon Trust's calculations of per capita carbon dioxide emissions are based not on production (the nationally emitted carbon dioxide divided by population), but rather on consumption (tracking the emissions of all goods consumed in the UK), categorized according to 'high-level consumer need' (Carbon Trust, 2006, p. 1). A consumption focus highlights the environmental impact of food and other consumer goods and services produced overseas, which are commonly excluded from these calculations, and in turn suggests a different set of carbon reduction policies from the government's focus on household energy use and transport. In turn, this draws greater attention to the consumption patterns of individuals, households and communities, and the scope for reducing carbon emissions from everyday activities and routine consumption patterns.

The scale of these challenges must not be underestimated; what is required is fresh thinking and new approaches to the economy–environment interface, to promote radical action. The Sustainable Development Commission of the UK government has recently published a report, *Prosperity Without Growth*, which sets out such an alternative agenda, encapsulating what has become known as a 'new economics' approach to sustainable development (Jackson, 2009). This manifesto for change in thinking and policy around economic policy argues that governments should eschew goals of economic growth in favour of promoting sustainable wellbeing within environmental limits. This view is broadly representative of an alternative theoretical approach to environmental governance and sustainable consumption known as the 'New Economics' (Daly and Cobb, 1990; Boyle, 1993; Robertson, 1999; Seyfang, 2009). This environmental philosophical and political movement stresses the benefits of decentralized social and economic organization and local self-reliance in order to protect local environments and economies from the negative impacts of globalization (Schumacher, 1973; Jacobs, 1984). Although its traditions go back much further (see Lutz, 1999), the UK's New Economics Foundation was established in 1986 to promote these ideas in research and policy (Ekins, 1986), for instance by developing new measures of wellbeing, seeking to understand consumer motivations in social context, and debating how an 'alternative' sustainable economy and society might operate.

By suggesting that societal systems of provision be examined, redesigned and reconfigured in line with sustainable consumption goals, the New Economics proposes nothing less than a paradigm shift for the economy. Rather than making incremental changes, the model entails a systemic change in economic and social infrastructure, altering the rules of the game and the objective of economic development. The New Economics therefore presents many challenges to mainstream thought and practice on sustainable consumption. Its objectives are to develop a practical approach to sustainable development which encompasses new definitions of wealth and work, new uses of money and which integrates ethics into economic life, and thereby to provide ecological citizens with the means to express their values and reduce their ecological footprints (Boyle, 1993; Wackernagel and Rees, 1996; Dobson, 2003).

When considering how to green consumption, this view rejects the orthodox 'ecological modernization' policy approach of providing information to consumers and relying on market transformation and greening (Maniates, 2002; Defra, 2003; Seyfang, 2005). Instead it acknowledges the important social and psychological aspects of consumption decisions, and the structural factors which result in much consumption being routine, habitual and effectively unconscious (Shove, 2003; Southerton et al., 2004; Jackson, 2007). Indeed, the new concept of 'carbon capability', which might be seen as a prerequisite for low carbon communities, stipulates that awareness of the factors outside one's own control is an important and empowering aspect of carbon management on an individual level, and one which prompts the move towards collective action and problem solving (Whitmarsh et al., 2009). Consequently, moves towards more sustainable consumption and lower carbon communities must therefore look beyond incremental improvements in efficiency, and instead aim for structural realignment in social institutions and systemic changes in provisioning (Seyfang, 2009).

Integrating these basic principles of redirecting economic development, cutting economic growth and promoting social and ecological sustainability, a New Economics strategy for sustainable consumption would therefore embody the following five priorities:

- *Localization*: strengthening local economies can occur through increasing the economic multiplier (the number of times money changes hands before leaving an area), which in turn occurs as a by-product of import substitution or local provisioning.
- *Reducing ecological footprints*: cutting material consumption and waste levels can be achieved through recycling, changing consumption patterns to cut demand, sharing facilities and resources, and so on.

- *Community building*: sustainable communities are robust, resilient, inclusive and diverse. Overcoming social exclusion, nurturing social capital and developing active citizenship within participative communities are key aspects of this.
- *Collective action*: this covers both acting collectively to influence decisions and deliver services, and also addressing questions of institutional consumption.
- *Building new social institutions*: creating new social and economic institutions – alternative systems of provision – which are based upon different conceptions of wealth, progress, value, and so on, and through these allowing people to behave as ecological citizens.

These five indicators of sustainable consumption are used below to discuss the aims and potential of complementary currencies, a range of new tools and instruments for building sustainable, low carbon communities.

INTRODUCING COMPLEMENTARY CURRENCIES

Faced with the challenge of forging new social institutions and systems of provision to deliver sustainable consumption, 'complementary currencies' have been proposed as a potentially useful tool. These are financial mechanisms which operate alongside conventional money, facilitating the exchange of goods and services in a parallel market, where alternative rules and resources prevail. Far from being exotic rarities, these instruments are in fact common features of modern life in developed countries: Airmiles is one such complementary currency – and reportedly the second largest currency in the world (BBC, 2002) – albeit one run by corporations to promote air travel and consumption of other material goods. The complementary currencies examined here are somewhat different; they emerge from the grassroots, or from civil society organizations, and pursue sustainability and social justice agendas (see also Seyfang, 2006b). What they have in common with Airmiles is that they are principally media of exchange, voluntarily accepted in exchange for goods and services. The conventional definition of money describes a tool which serves three economic purposes: as a medium of exchange, a store of value and a unit of account (Lipsey and Harbury, 1992). However, it is only in recent times that these functions have been combined within a single monetary instrument, known as general purpose money.

Furthermore, as Keynes (1936) observed, these functions can conflict with each other, as is seen in recession when money's function as a store of value results in it being hoarded rather than circulated, and the supply of money dries up, preventing exchange. Complementary currencies can be regarded as 'special purpose money' as they tend to fulfil one or other of money's functions to the exclusion of others, for instance offering a medium of exchange which is incentivized to encourage monetary circulation rather than hoarding (Seyfang, 2000). Indeed, history shows that economic downturns are often accompanied by an upsurge in interest in these tools for local economic resilience (Douthwaite, 1996; Boyle, 2002).

Money is a socially constructed tool, and its design imbues it with particular behavioural incentives. Lietaer states, 'Money matters. The way money is created and administered in a given society makes a deep impression on values and relationships within that society. More specifically, the type of currency used in a society encourages – or discourages – specific emotions or behaviour patterns' (Lietaer, 2001, p. 4). There are many diverse models of complementary currency in use around the world, from Japan and Canada to Mexico, Malaysia, Europe and Argentina (www.complementarycurrency.org). The form and design of these currencies reflect the contexts within which they are developed and the purposes intended for their use. For instance, the London-based 'Wedge' card is a loyalty scheme for customers of local independent businesses, to promote more local shopping (www.wedgecard.co.uk). In the Netherlands, plans for an Amsterdam City Carbon Card are being developed, to incentivize consumption of more sustainable products while also rewarding community involvement activities (www.qoin.com).

In Japan, the demographics of aging (soon to be experienced in the UK) have resulted in social care needs exceeding the state's ability to provide; reciprocal social care schemes have developed such as 'fureai kippu', allowing carers to earn points helping elderly people, which can be spent on receiving care later on or donated to others (this system echoes the time banks model below). In other areas such as Hungary, Canada, Honduras and South Africa, currencies have been developed to supplement scarce national currency and enable local trade, and in the UK the most common complementary currency is LETS (Local Exchange Trading Schemes), which had around 300 projects in the late 1990s (Seyfang, 2001; Williams et al., 2001). In Perth, Australia, an experimental greenhouse gas reduction currency called the Maia Maia Project aims to reward participants who commit to reducing their carbon emissions with a currency convertible to carbon dioxide-equivalent global warming potential, which attracts discounts in local businesses (themaia-maiaproject.blogspot.com).

The complementary currencies examined here are designed for specific aims concerning social, economic and environmental aspects of sustainable development, and in particular the need to transform the consumption patterns of individuals, households and communities. The following section introduces two distinct types of complementary currency which are in use in the UK today (time banks and local money), and describes their origins and characteristics. It then proceeds according to the criteria of sustainable consumption set out above, and highlights where these complementary currencies are attempting to make a difference to the financial infrastructure of society, and the implications of those changes for low carbon communities.

CURRENCIES OF CHANGE

Time Banks: Spending Time Building Community

Time banking is a social justice movement which prioritizes social care, wellbeing and reciprocity. After LETS, it represents the second wave of complementary currencies in the UK, and is based on the US time dollar model developed by Edgar Cahn, which aims to rebuild supportive community networks of reciprocal self-help, particularly in deprived neighbourhoods. A time bank is essentially a volunteering exchange, with a central broker to coordinate members' activities. Everyone's time is worth the same – one time credit per hour – regardless of the service provided. Participants earn credits by helping others and spend credits receiving help themselves (Cahn and Rowe, 1992).

The first UK time bank was set up in 1998, and in 2002 there were 36 active time banks, with 2196 participants in total and nearly 64,000 hours exchanged (Seyfang and Smith, 2002). There are currently 109 in operation and a further 48 being developed (Ryan-Collins et al., 2008). Time banks aim to overcome the 'green niche' limitations of the previous wave of complementary currencies (LETS) by being based in mainstream institutions (health centres, schools, libraries), paying coordinators for development and support work, and, most importantly, for brokering transactions between participants (Seyfang, 2002), but they still face obstacles in achieving their potential. These are: large 'skills gaps' in projects, which again presents a limited range of services available; short term funding mitigates against projects which take a long time to become established (annual project costs were estimated to be £27,300 in 2002); reciprocity is slow to materialize due to the reluctance of participants to ask for help; and while the unemployed are officially encouraged to participate in time banking, those in receipt of disability benefits face particular obstacles from the benefit system (Seyfang and Smith, 2002; Seyfang, 2003, 2004a,b).

Turning to the criteria of sustainable consumption set out above, it is clear that time banks have the potential to impact on the consumption patterns of participants in several ways, both direct and indirect. The services provided on a time bank – neighbourly support such as dog walking, gardening, small DIY tasks, and so on – tend to be locally based by definition, so there is no net localization effect; the time bank creates new local networks and opportunities for exchange, and does not substitute for imports. Similarly, although reducing environmental impact is not a primary objective, time banking is being used to promote more sustainable consumption and environmental governance in a variety of ways. In north London, for example, residents of an inner city estate will soon be able to earn time credits for recycling their household waste and spend them on attending training courses or refurbished computers. Another London time bank rewards members with low energy light bulbs. Participation in groups which make local environmental decisions could also be rewarded. Furthermore, it might be expected that by providing participants with access to supportive networks, where resources and expertise can be shared, some material consumption might be avoided. Examples of this type of environmental impact were found in previous studies of LETS, where members hired tools and so avoided having to purchase their own (Seyfang, 2001).

As indicated above, the primary rationale for time banking is community building, and the projects are successful at developing social capital and new supportive networks. They attract members of the most socially excluded groups in society (those who normally volunteer least), and are often introduced into marginalized areas where building trust and neighbourliness is a challenge which the conventional economy cannot meet. For socially excluded individuals and communities, whose skills are accorded no value in the mainstream economy, the opportunity to be valued and rewarded for one's input into community activity and for helping neighbours is enormously empowering. There is also a collective action aspect to time banking. In addition to the 'community time bank' model, time banks can be used as a 'co-production' tool to encourage people to become involved in the delivery of public services which require the active participation of service users in order to be successful, for example health, education, waste management, local democracy, and so on (Cahn, 2000; Ryan-Collins et al., 2008) and 'co-production is a framework with the potential for institutions . . . to achieve the elusive goal of fundamental and systemic change' (Burns, 2004, p. 18).

By rewarding and encouraging civic engagement, time banks could invigorate active citizenship. Finally, the most significant benefit of time banking, for many participants, is the opportunity to redefine what is considered 'valuable', in other words creating new institutions of wealth, value and work (Seyfang, 2004a,b). The radical stance of valuing all labour (or time) equally seeks to explicitly recognize and value the unpaid time that people spend maintaining their neighbourhoods and caring for others. Thus voluntary work is rewarded and so incentivized (rather than squeezed out by the conventional economic system which accords it no value and so undermines social cohesion), thereby ensuring that vital socially reproductive work is valued and carried out (Seyfang, 2006a). Time banks represent a new infrastructure of income distribution for society, where income is not dependent upon one's value to, and activity in, the formal economy, but rather upon work – broadly defined (Burns, 2004; Seyfang, 2006a).

In addition to the positive social inclusion impacts of time banks and beyond the immediate impacts on consumption discussed above, time banks offer further possibilities for shifting consumption patterns towards lower carbon options. Given that the principal aim and impact of time banking is to address unmet needs and offer recognition for community engagement, there are strong social and psychological impacts resulting from participation. These include feeling acknowledged, gaining self-esteem, demonstrating belongingness, expressing values and accessing personal contact.

All of these deep-rooted needs are among the range of social and psychological drivers of consumption. We often consume material goods to meet these non-material needs, and this has profound implications for sustainable development and the drive to cut carbon emissions (Max-Neef, 1992; Jackson, 2007). Consumption is used to boost esteem, to feel part of a community, to express ourselves and to connect with others – even when this consumption undermines other needs such as personal safety or ecological sustainability. As time banks deliver some of these social and psychological benefits through a structure of supportive social networking, it might be expected that participants are therefore less likely to seek to meet those needs through material consumption, so lowering their overall consumption levels and associated carbon emissions. This is an empirical question for future research on time banks, and could offer important lessons on how consumers may be weaned off their material consumption fix, without compromising needs-satisfaction.

Local Money Systems: Supporting Local Businesses for Economic Resilience

Across the UK, there are examples of communities which are printing their own money and spending it in local businesses, knowing that it must circulate locally and stay in the area, rather than leaking away as conventional money is wont to do. Local money systems such as the Totnes Pound (totnes.transitionnetwork.org/totnespound/home) and the Lewes Pound (www.thelewespound.org) represent the third wave of complementary currencies in the UK. They build on previous experiences with LETS and international experience in local money systems, such as the US Berkshares (www.berkshares.org) and Saltspring Dollar schemes (www. saltspringdollars.com) and German Cheimgauer (www.chiemgauer.info), part of that country's 'regiogelder' or regional money movement (www. regiogeld.de), which currently has 28 active currencies (Gelleri, 2009). This money is legal as a voucher but not as legal tender (that is, no-one is compelled to accept it, and taxes on local money income must be paid in national currency). These are asset-backed currencies (members of the public and businesses exchange Sterling for local Pounds), and local businesses choose to accept them as part-payment for goods and services; individuals can also exchange them between themselves.

In contrast to time banking, with its principally social objective, local money systems aim to strengthen local economies to offer a greater degree of resilience against economic shocks and support local exchange in times of economic downturn. In this, local money echoes the goals of the earlier wave of LETS, but the mechanism and design of the currency are somewhat different, and it aims to appeal to a broader cross-section of the population. For instance, the Totnes Pound in Devon, the first of this type in the UK, was launched in March 2007, with a deliberate aim to diversify and experiment with new types of complementary currency in the UK. Following from this, Lewes in Sussex launched its own currency in September 2008.

Both these UK initiatives have grown from groups of community activists seeking to develop a more sustainable money system. Indeed, they have emerged from local Transition Town groups, which have the stated aim of mobilizing grassroots action to move towards low carbon, resilient communities. As the Lewes Pound website states, the local currency 'benefits shoppers by creating stronger and more local shops, increasing a sense of pride in our community, decreasing CO_2 emissions and increasing economic resilience' (Transition Town Lewes, 2008). How successful have they been at achieving these aims? Initial reports in the media and on practitioner websites suggest that their impact has so far been more symbolic than economic, and both projects are considering how to evolve and develop in the future to learn from initial experiences (some details of which formed the focus of a recent edition of *You And Yours* broadcast on radio by the BBC, 2009). To date there has been no published empirical research on these experiments, which renders a critical evaluation problematic, but indicative impacts can be suggested using the criteria for sustainable consumption outlined above.

Unlike conventional money, the fundamental characteristic of local money is that it can only be spent within a defined local or regional area. Thus, Lewes Pounds can only be spent among individuals and businesses in Lewes, and they must keep circulating (increasing the local multiplier) until they are converted back to Sterling. In this way, the aim is to strengthen local economic linkages and promote import substitution, thereby nurturing local economic resilience, as opposed to dependence on external markets, and reducing transport costs (and carbon emissions) from imports. This objective speaks directly to the New Economics goals of localization of economic systems and reducing ecological footprints. While the impacts of these projects are currently unresearched, the Cheimgauer regional money system in Germany is comparable and might indicate possible future impacts.[1] This currency launched in 2003, and Gelleri estimated that the velocity of circulation of the local currency by 2008 was three times that of the Euro, and that approximately 2500 consumers exchanged €100,000 into Cheimgauer notes each month, which they could spend at 600 local businesses. In 2008 the turnover increased by 30 per cent over the previous year, to €3.6 million, and an estimated one third of this represents new economic activity which would not have occurred before. This includes significant import-substitution benefits such as local biofuels replacing imported oil (Gelleri, 2009). Of course, the overall impact on sustainable consumption depends on the extent to which this new consumption is different from pre-existing patterns, rather than more of the same, and whether it is in addition to, or replaces, that consumption.

The community-building aspects of these local currency notes is significant, perhaps eclipsing the economic impacts in these early stages, as was previously found with LETS (Seyfang, 2001). While electronic currencies are more efficient, it is certainly true to say that the physical notes themselves in a printed currency hold great symbolic value for local residents. High quality, counterfeit-proof paper displaying local landmarks or notable citizens are a concrete expression of local pride and solidarity. The Lewes Pound, for instance, pictures local hero Thomas Paine, and carries his quotation 'We have it in our power to build the world anew' (www.thelewespound.org). By presenting the currency as a token of positive local action, it aims to appeal to a wide cross-section of the population. Furthermore, by encouraging consumers to favour local businesses, there is a positive benefit in terms of building supportive links between townspeople and their traders. Other than this, however, there is little scope at present for collective action to be generated directly through the use of a local currency system. Essentially the currency is an individualistic tool, to complement existing transactions in national currency, rather than a structure to generate collectivity. However, despite this, the currency strongly represents a new financial institution or system of provision. Its rules, mechanisms and practices differ from mainstream exchange mechanisms in their local-boundedness, and thus they promote a different set of ideas and practices about economic behaviour.

Experience to date with the Totnes and Lewes currencies indicates that while some people and businesses use the local money enthusiastically, other businesses find it difficult to spend the notes, and so the economic impact is constrained (BBC, 2009). In the USA, similar projects overcome this barrier by paying some of their staff's wages in local currency, and it may be that as the currencies grow in scale and achieve a critical mass in a town, they can become ubiquitous and easier to earn and spend. In the UK, the Lewes and Totnes Pound groups are also investigating the possibility of introducing electronic versions of their local currency, to more closely match the transaction style of businesses and consumers alike. These nascent projects demand further empirical study to investigate their economic impacts and the ways in which the new currencies mediate consumption.

CONCLUSIONS: GROWING GRASSROOTS INNOVATIONS

Sustainable consumption requires a radical realignment of social and economic institutions and systems of provision. One such system is that of exchange. The two types of complementary currency examined here each present distinct possibilities for reducing carbon emissions and moving towards low carbon communities, while simultaneously meeting some or all of the New Economics criteria for sustainable consumption. Time banking appears to offer the greatest potential for carbon reduction through offering a supportive social network which meets some of the participants' social and psychological needs for recognition, esteem and belongingness – needs which might otherwise be met through material consumption. A secondary impact is through the potential for sharing resources. In contrast, local money systems aim to strengthen and build resilience in local economies, and so their principal impact on consumption is through localization and import substitution – which brings carbon reductions from avoiding transport costs. What the two initiatives share is a commitment to building new systems of exchange which meet sustainability criteria and express a wider set of values than mainstream money.

In this, they can be seen to be successful, albeit on a small scale. However, their potential for wider impact is much greater. The 'sustainability transitions' literature describes the role of experimental niche projects in seeding wider system change (Geels, 2002; Schot and Geels, 2008). While this literature is mainly focused on technological innovations in commercial organizations, an emerging 'grassroots innovations' literature is exploring the applicability of these models for social innovation in the social economy (Seyfang and Smith, 2007; Seyfang, 2009). From this perspective, complementary currencies can be viewed as social innovations in exchange. Their interactions with mainstream exchange and regulatory systems, as they attempt to replicate and grow, are complex and problematic – not least because the projects' values differ markedly from those of the mainstream economic system (see Seyfang, 2006a; Smith, 2007).

Despite official support, contradictory and occasionally unhelpful policies can block the wider adoption of projects like this, as the time banking example demonstrates. But there are further institutional issues which hinder the development of alternative green niches of this type. Commonly, these are a need for funding to support new experimental initiatives and test new practices, a lack of institutional learning (scarce time is spent securing funding rather than developing learning mechanisms and consolidating knowledge), and the need for networking institutions to support initiatives, share best practice and liaise with partnering organizations, government and business to translate ideas into new settings (Smith, 2007). In the cases of these complementary currencies, efforts have been made to address these issues, for instance through international and national time-banking organizations (www.timebanks.org; www.time-banking.org), and through conferences on local and regional monetary systems across Europe (www.regiogeldkongress.de).

Nevertheless, there is scope for much greater policy support, and it may be that lessons can be drawn from the more conventional literature on supporting and diffusing innovation. For example, could incubators for social innovations bring benefits from clustering and critical mass? What is the scope for 'policy test sites' where normal regulatory rules are relaxed to allow experimentation, testing and failure? What happens when small scale projects grow and become translated to more mainstream settings? How can these processes be managed without alienating the value-driven activists who instigated them as alternatives to the mainstream? These dynamic diffusion processes are the subject of ongoing research, and it is hoped that future studies will provide a greater understanding of the ways in which initiatives of this type can be supported and harnessed to achieve wider policy goals for low carbon communities.

ACKNOWLEDGEMENTS

The author would like to thank the UK's Economic and Social Research Council and Research Councils UK for funding the research on which this chapter is based (through project R000223453, CSERGE Centre funding and an Academic Fellowship in Low Carbon Lifestyles). Thanks also to Tom Hargreaves and the book's editors for helpful comments on an earlier version of the chapter.

NOTE

1. The Cheimgauer differs in one important way from the UK local currencies: it slowly devalues over time. This 'demurrage' or negative interest fee of 2 per cent a quarter must be paid by people holding the notes, before they can be spent. This incentivizes rapid circulation and discourages hoarding (Gelleri, 2009). The mechanism is being considered by the UK projects as part of their evolution and development.

REFERENCES

BBC (2002), 'Air Miles "threaten dollar's dominance"', *Business News*, 3 May, accessed at http://news.bbc.co.uk/1/hi/business/1966290.stm.
BBC (2009), *You And Yours*, Radio 4, 16 March.
Boyle, D. (1993), *What Is New Economics?*, London: New Economics Foundation.
Boyle, D. (ed.) (2002), *The Money Changers: Currency Reform From Aristotle To E-Cash*, London: Earthscan Ltd.
Burns, S. (2004), *Exploring Co-production: An Overview of Past, Present and Future*, London: New Economics Foundation.
Cahn, E. (2000), *No More Throwaway People: The Co-production Imperative*, Washington, DC: Essential Books.
Cahn, E. and J. Rowe (1992), *Time Dollars*, 2nd edn, Chicago, IL: Family Resource Coalition of America.
Carbon Trust (2006), *The Carbon Emissions Generated In All That We Consume*, London: Carbon Trust.
Daly, H. and J. Cobb (1990), *For the Common Good*, London: Greenprint Press.
Defra (Department for Environment, Food and Rural Affairs) (2003), *Changing Patterns: UK Government Framework for Sustainable Consumption and Production*, London: Defra.
Dobson, A. (2003), *Citizenship And The Environment*, Oxford: Oxford University Press.
Douthwaite, R. (1996), *Short Circuit: Strengthening Local Economies for Security in an Unstable World*, Totnes, UK: Green Books.
Druckman, A., P. Bradley, E. Papathanasopoulou and T. Jackson (2008), 'Measuring progress towards carbon reduction in the UK', *Ecological Economics*, **66** (4), 594–604.

Ekins, P. (ed.) (1986), *The Living Economy: A New Economics in the Making*, London: Routledge.

Geels, F.W. (2002), *Understanding the Dynamics of Technological Transitions*, Enschede, Netherlands: Twente University Press.

Gelleri, C. (2009), 'Chiemgauer regiomoney: theory and practice of a local currency', *International Journal of Community Currency Research*, **13**, 61–75, accessed at www.uea.ac.uk/env/ijccr/.

HM Government (2008), *Climate Change Act 2008*, London: The Stationery Office.

IPCC (Intergovernmental Panel on Climate Change) (2007), *Climate Change 2007: Synthesis Report. Contribution of Working Groups I, II and III to the Fourth Assessment Report of the Intergovernmental Panel on Climate Change*, core writing team R.K. Pachauri and A. Reisinger (eds), Geneva: IPCC.

Jackson, T. (2007), 'Consuming paradise? Towards a social and cultural psychology of sustainable consumption', in T. Jackson (ed.), *The Earthscan Reader In Sustainable Consumption*, London: Earthscan, pp. 367–95.

Jackson, T. (2009), *Prosperity Without Growth? The Transition to a Sustainable Economy*, London: Sustainable Development Commission.

Jacobs, J. (1984), *Cities And The Wealth Of Nations: Principles of Economic Life*, London: Random House.

Keynes, J.M. (1936 [1973]), *The General Theory Of Employment, Interest And Money*, London: Macmillan.

Lietaer, B. (2001), *The Future Of Money: Creating New Wealth, Work and a Wiser World*, London: Century.

Lipsey, R. and C. Harbury (1992), *First Principles of Economics*, 2nd edn, Oxford: Oxford University Press.

Lutz, M. (1999), *Economics For The Common Good: Two Centuries of Social Economic Thought in the Humanistic Tradition*, London: Routledge.

Maniates, M. (2002), 'Individualization: plant a tree, buy a bike, save the world?', in T. Princen, M. Maniates and K. Konca (eds), *Confronting Consumption*, London: MIT Press, pp. 43–66.

Max-Neef, M. (1992), 'Development and human needs', in P. Ekins and M. Max-Neef (eds), *Real-Life Economics: Understanding Wealth Creation*, London: Routledge.

Robertson, J. (1999), *The New Economics Of Sustainable Development: A Briefing For Policymakers*, London: Kogan Page.

Ryan-Collins, J., L. Stephens and A. Coote (2008), *The New Wealth Of Time: How Time Banking Helps People Build Better Public Services*, London: New Economics Foundation.

Schot, J. and F. Geels (2008), 'Strategic niche management and sustainable innovation journeys: theory, findings, research agenda, and policy', *Technology Analysis and Strategic Management*, **20** (5), 537–54.

Schumacher, E.F. (1973 [1993]), *Small Is Beautiful: A Study Of Economics As If People Mattered*, London: Vintage.

Seyfang, G. (2000), 'The euro, the pound and the shell in our pockets: rationales for complementary currencies in a global economy', *New Political Economy*, **5** (2), 227–46.

Seyfang, G. (2001), 'Community currencies: small change for a green economy', *Environment and Planning A*, **33** (6), 975–96.

Seyfang, G. (2002), 'Tackling social exclusion with community currencies:

learning from LETS to time banks', *International Journal of Community Currency Research*, **6**, accessed at www.uea.ac.uk/env/ijccr/.

Seyfang, G. (2003), 'Growing cohesive communities, one favour at a time: social exclusion, active citizenship and time banks', *International Journal of Urban and Regional Research*, **27** (3), 699–706.

Seyfang, G. (2004a), 'Time banks: rewarding community self-help in the inner city?', *Community Development Journal*, **39** (1), 62–71.

Seyfang, G. (2004b), 'Working outside the box: community currencies, time banks and social inclusion', *Journal of Social Policy*, **33** (1), 49–71.

Seyfang, G. (2005), 'Shopping for sustainability: can sustainable consumption promote ecological citizenship?', *Environmental Politics*, **14** (2), 290–306.

Seyfang, G. (2006a), 'Harnessing the potential of the social economy? Time banks and UK public policy', *International Journal of Sociology and Social Policy*, **26** (9/10), 430–43.

Seyfang, G. (2006b), 'New institutions for sustainable consumption: an evaluation of community currencies', *Regional Studies*, **40** (7), 781–91.

Seyfang, G. (2009), *The New Economics of Sustainable Consumption: Seeds of Change*, Basingstoke: Palgrave Macmillan.

Seyfang, G. and A. Smith (2007), 'Grassroots innovations for sustainable development: towards a new research and policy agenda', *Environmental Politics*, **16** (4), 584–603.

Seyfang, G. and K. Smith (2002), *The Time of Our Lives: Using Time Banking for Neighbourhood Renewal and Community Capacity-Building*, London: New Economics Foundation.

Shove, E. (2003), *Comfort, Cleanliness and Convenience: The Social Organization of Normality*, Oxford: Berg Publishers.

Smith, A. (2007), 'Translating sustainabilities between green niches and socio-technical regimes', *Technology Analysis and Strategic Management*, **9** (4), 427–50.

Southerton, D., H. Chappells and V. Van Vliet (2004), *Sustainable Consumption: The Implications of Changing Infrastructures of Provision*, Cheltenham, UK and Northampton, MA, USA: Edward Elgar Publishing.

Transition Town Lewes (2008), *The Lewes Pound: What Is It?*, accessed 8 April 2009 at www.thelewespound.org/what.html.

Wackernagel, M. and W. Rees (1996), *Our Ecological Footprint: Reducing Human Impact on the Earth*, Philadelphia, PA: New Society Publishers.

Whitmarsh, L., S. O'Neill, G. Seyfang and I. Lorenzoni (forthcoming), 'Carbon capability: understanding, ability and motivation for reducing carbon emissions', in A. Stibbe (ed.), *The Handbook of Sustainability Literacy: Skills for Surviving and Thriving In the Twentyfirst Century*, London: Green Books.

Williams, C.C., T. Aldridge, J. Tooke, R. Lee, A. Leyshon and N. Thrift (2001), *Bridges into Work: An Evaluation of Local Exchange Trading Schemes (LETS)*, Bristol: Policy Press.

8. Decarbonizing local economies: a new low carbon, high well-being model of local economic development

Elizabeth Cox and Victoria Johnson

INTRODUCTION

Over the next 10 years the UK will have to have responded to the challenge of building a new economic system, not only at the global and national level, but most importantly to articulate a framework of outcomes and actions required at the regional and sub-regional levels that will support more sustainable lifestyles.

At the heart of our approach is the contention that the limits to adaptation are endogenous to society, a reflection of societal values, which while they can evolve and change over time, are not common across all groups in society (Adger et al., 2008). Any process of local adaptation needs to start by making explicit these diverse social values, the connection between decisions taken by individuals, households and institutions, and the outcomes created by these decisions. The desirability of these outcomes can then be tested against a set of wider societal goals. Forms of organizational structure and governance, and the scale and scope of delivery would then have to be judged in reference to their fitness to deliver these wider goals. Understanding that it is possible to change, or re-engineer, the economy to better serve societal goals at the national and local level is a first and necessary step to decarbonizing local economies.

This chapter is exploratory and aims to provoke debate regarding the purpose of an economy, the required benefits flowing from it, and the constituent elements of an effective approach to support communities to make the transition to low carbon lifestyles.

THE NEED FOR A RAPID AND JUST TRANSITION

Research correlating the latest carbon emission trends with the sensitivity of the climate system has suggested that from August 2008 there may be only 100 months before we enter a potentially irreversible phase of climate change (Anderson and Bows, 2008; Johnson and Simms, 2008). Failure to meet the challenge by transitioning to a lower carbon economy within this relatively short time period could leave the coming decades characterized by a series of self-reinforcing social, economic and ecological crises. The existence of tipping points in the climate system and unpredictable environmental feedback-loops also mean that change and impacts are likely to be non-linear (Rial et al., 2004; Lenton et al., 2008). It then follows that the core policy responses, both mitigating against and adapting to climate change, will in turn need to be non-linear in character. Typically, any attempt at incremental change will be insufficient to keep pace with environmental change. What is required is a systematic response which is mutually reinforcing at the community, institutional, macro and global levels (that is, across the development context).

Adaptation and Resilience

Combating climate change locally will require both adaptation and resilience characteristics to be designed into the structure of the local economy. Adaptation to climate change or resource scarcity can be thought of as the long-term increase in the capacity to cope with changes in circumstances (for example, unforeseen or periodic hazardous events). Adaptation at the local level will require the 'climate proofing' of infrastructure, investments and activities.

There has been a growing recognition in climate change literature over the last decade that the capacity to adapt cannot be reduced to a discussion of vulnerability to the direct impacts of climate change (for example, flooding and heatwaves), but that it also includes political, economic and social factors (Adger et al., 2006; Smit and Wandel, 2006). Socially constructed limits to adaptation have been described by Adger et al. (2008) as comprising values (what is important to society), knowledge (how and what we know), risk (how and what we perceive) and culture (how we live). Whilst all these socially constructed limits can be overcome, the implication is that each of these limits will have to be systematically addressed in any approach that supports communities to develop their adaptive capacities.

Resilience can broadly be defined as the ability of a system (social, economic or ecological) to cope with external shocks as they arise. In measuring a system's resilience, the Tyndall Centre for Climate Change Research refers to indicators that demonstrate the system's ability to (a) absorb shocks and retain its basic function, (b) self-organize (social institutions and networks), and (c) innovate and learn in the face of disturbances (Adger et al., 2004). Supporting the development of resilient qualities in economic and social systems is particularly important when there are high levels of uncertainty regarding the direct and indirect impacts of climate change and resource scarcity.

Recent research by Bartley (2008) explored why some 'deprived' communities are more resilient than others to external shocks, for example a sudden loss of employment opportunities. The findings suggest that it is social capital that holds a community together in such circumstances. Of the various definitions of social capital, Edwards et al. (2003, p. 2) offer one that maps easily onto discussions of well-being explored later in this chapter, defining it as:

> the values that people hold and the resources that they can access, which both result in and are the result of collective and socially negotiated ties and relationships. The extent to which people share a sense of identity, hold similar values, trust each other and reciprocally do things for each other, then this is felt to have an impact on the social, political and economic nature of the society in which they live.

Cahn (2000), in discussing the core economy, offers a broader insight into those aspects of a community which support resilience. The core economy is made up of human resources embedded in the everyday lives of individuals (time, wisdom, experience, energy, knowledge, skills) and in the relationships between individuals (love, empathy, reciprocity, teaching, learning). Cahn argues that it is the core economy which underpins the market economy, through activities such as raising children, caring for people, feeding families, maintaining households, and building friendships, social networks and civil society. These are largely un-commodified, unpriced and unpaid functions (Stephens et al., 2008) which are essential to supporting resilience at the individual and community level. The resilience of individuals is also found to be affected by their most significant relationships, types of schooling or training available, job opportunities, and the social and legal barriers which limit choices (Bartley, 2008).

Overall, this implies that in order to create the conditions for successful adaptation and enhanced resilience which create socially just outcomes at the national to local level, it is necessary to redress inequalities such as income, health and education, in addition to supporting growth of the core economy thorough action at the institutional and macro levels.

GOALS OF A NEW ECONOMY

The values of a society express what is important and drive how scarce resources are allocated and controlled through rules, and organizational and governance structures (Adger et al., 2008). That is, they underpin the design of the economy. We argue in this chapter that the purpose of the economy should be to enhance the well-being of UK citizens in a way that is socially just and environmentally sustainable (Cox, 2008).

At the local level this revised economic model will require: a fundamental re-engineering of the local economy to achieve its new purpose; a reassessment of how progress is measured and who measures it to provide the necessary feedback loops to support adaptation; development of new institutional structures and supportive policies at the national and sub-national levels; and an understanding of how change can be mobilized at the level most immediately felt by individuals, families and communities through the environment on their doorsteps, the goods and services that they produce and consume, and their sense of well-being. The defining principles of the new model are explored below.

Well-Being

Well-being is a dynamic process of mutually reinforcing feedback loops, resulting from the way in which people interact with the world around them. Subjective well-being studies suggest that, in addition to experiencing good feelings, well-being involves things such as participating in activities which are meaningful and engaging, which support individuals to feel competent and autonomous. Well-being also involves individuals being able to draw on a stock of inner resources to help support a feeling of resilience in the face of changes they cannot control; a sense of individual drive and energy (vitality); a sense of connectedness to other people, and being able to draw upon supportive relationships (Michaelson et al., 2009).

In the context of adaptation (both community and at the individual level), high levels of well-being imply that we are more able to respond to difficult circumstances, innovate and engage with people around us.

Sustainable Social Justice

The concepts of social justice and environmental sustainability are combined by Coote and Franklin (2009, p. 5) when they define sustainable social justice as 'the fair and equitable distribution of social, environmental and economic resources between people, countries and generations'.

Applied to the local economy, sustainable social justice implies equality of access and control of resources, a raised level of awareness and understanding about choices made, and a shared responsibility within and between generations, which can be applied across all three resource areas. Equal access and control refers to opportunities (for example, work, education), governance structures (for example, mutuality) and deliberative processes that allow all voices to be heard in decision-making processes (for example, consensus voting, participatory budgeting). A shared responsibility within a community and to future generations refers to personal responsibility for decisions made, and community-level responsibility for neighbours is best exemplified through the core economy in the provision of safe, vibrant neighbourhoods, democracy, love and care (Stephens et al., 2008). All of which have the potential to enhance the physical and social conditions around us, directly affecting our abilities to function well and, in the case of the strongest links, building our psychological resources to support positive emotions. Positive emotions are known to help bolster psychological resilience to enable children and families to better cope with difficult circumstances (Shannon et al., 2006; Hughes and Kendall, 2008; Reschly et al., 2008).

Low Carbon

The overall aim of a low carbon society, working on scales from the local to the national, should be to live within the limits of the planet's biocapacity; that is, one planet living. Existing infrastructure and institutions operating at spatial scales larger than the local level can act as barriers to rapid change. Urban design, energy and transport systems, and buildings, once in place, tend to 'lock in' patterns of resource use and consumption that last for decades (Unruh, 2000). Incremental change achieved by substituting improved alternatives is unlikely to match the scale or urgency of the challenge. It has been estimated that over 80 per cent of the CO_2 released into the atmosphere from human activity is due to the burning of fossil fuels, mainly used to produce electricity for heat or for transport. Economic and population growth also mean that global energy demand is accelerating (Raupach et al., 2007). Climate change and the decline of cheap, abundant oil and gas point to a need for a renewable-based energy system to replace fossil fuels as far as possible.

Demand reduction among 'over-consumers' is equally important. In the UK, in general, the poorest 10 per cent of households produce only 45 per cent as much CO_2 from their homes as the richest 10 per cent (Roberts, 2008).[1] Low income households that are also low emitters are more likely to be both fuel poor and employ coping strategies, such as not heating their homes (which often themselves exhibit poor thermal characteristics).

Functional Consumption

In order to reduce per capita carbon emissions to a sustainable level, there needs to be an absolute reduction in levels of consumption, given that a significant proportion of carbon emissions are related to the energy used in the production of consumer goods. Jackson (2005) argues that there are two key drivers of consumption in developed nations which are unrelated to the functional use of the goods and services consumed. The first relates to the symbolic role of consumer goods, whilst the second relates to the locking-in of consumers into unsustainable consumption patterns. This consumer lock-in inhibits consumers from exercising alternative consumption choices, and occurs in part through economic constraints, institutional barriers, inequalities in access and restricted choice. For example, rural populations can be locked into private transport use because of poor public transport provision. But consumer 'lock-in' also flows from habits, routines, social norms and expectations, and dominant cultural values (Sanne, 2002; Jackson, 2005).

Unequal patterns of consumption are also linked to issues of social justice. Socially excluded groups, in low income households in particular, spend a greater proportion of their income on energy and food. This is exacerbated by volatile global commodity prices and energy and food bills. Therefore policies which create the enabling conditions for a reduction in absolute consumption (for example, energy efficiency) or alternative systems of service provision that buffer against price volatility (for example, local food and renewable energy systems) provide an important social justice function.

It can be argued, therefore, that changing behaviour cannot be conceived as a process of simply encouraging change at the individual level. Pro-environmental behavioural change has to be a social process which needs policy support to effect larger scale change.

New Measures of Progress

Our current economic model in the UK places economic growth, measured as increases in GDP,[2] as the key national measure of progress. GDP is however a poor measure of societal progress, or adaptive capacity, as it can tell us little about quality of life, environmental costs or levels of existing inequality. The measure also excludes non-monetarized costs and benefits such as the value of unpaid work (Marks et al., 2006).

Authors such as Diener and Seligman (2004) make it clear that there are a number of factors – such as physical and mental health, family security, environmental quality and social cohesion – which contribute to well-being but which are not captured by conventional measures of economic output.

The goals of a new economy need to be supported by more meaningful measures of progress which capture the efficiency of resources used in producing long and satisfied lives.

SUPPORTING PRACTICAL ACTION IN COMMUNITIES

Over a four year period nef, in association with emda (East Midlands Development Agency), developed and piloted a new approach to local economic development[3] with communities experiencing economic disadvantage. The aim of the approach was to support increasing levels of community-level action which would, over a period of time, lead to the transformation of their local economies.

The elements of the approach included:

- An enterprising communities framework which explicitly described the characteristics of an enterprising community, and the desirable direction of change in terms of economic, social and environmental outcomes.
- A set of economic literacy tools that were used to engage community members in debate, raise awareness and understanding about the local economy, and supported the identification of opportunities for enterprise. Opportunities were identified through a process of developing understanding of the existing flows of money and resources into and out of their communities, and the passion for action.
- Supporting action through community-based support in the form of a coach and local networks.

The framework distilled the logic of the approach, describing an enterprising local economy as having seven mutually reinforcing characteristics:

- A responsible enterprise and business sector: a diverse range of businesses and enterprises in terms of size and social and private ownership, and diversity of goods and services produced.
- Positive local money and resource flows: high local multiplier in terms of spending and re-spending financial resources locally, and local re-use of waste, energy and resources.
- A strong local asset base defined as including local people's attitudes, skills and knowledge, and physical, financial and natural resources.
- A responsive public and business sector which is working to strengthen and invest in the local economy.

- A strong community and civic voice evidenced by local activism, leadership, volunteering and engagement in debate.
- Environmental sustainability and a reduced environmental footprint.
- Increased understanding of economic, cultural and ecological interconnections that link communities, span the globe and impact on the future (interdependence).

The process of engagement was focused on opportunities (rather than need identification) and started from a practical issue being faced by a particular community. Both of these were key motivators for involvement and action. Examples included the development of a new hospital in an area, local business concerns about a new by-pass, a local festival and having no available community meeting space. The coaching approach used to support the process focused individuals on what they were passionate about, supported them to recognize assets to which they had immediate access (skills, knowledge, physical resources), and drew on local networks of support to access other resources that were needed.

The defining characteristic of the economic literacy tools was that they were visually based tools which encouraged debate and exchange of information across different groups – particularly residents, business people, the third sector and the public sector. Within the community four basic questions were explored:

1. What are the opportunities for enterprise development of both new and existing businesses?
2. How could goods and services be delivered differently?
3. How can we mobilize resources to do what we want to do?
4. What are the local economic, social and environmental outcomes of our decisions?

Each of the pilot communities selected to be part of the programme had low reported levels of enterprise activity. Yet in each a positive enterprise response resulted from the process. The evaluation of the programme identified raised awareness of the local impact of spending patterns as a key impact.

SUPPORTING COMMUNITIES TO RE-ENGINEER THEIR LOCAL ECONOMY

Drawing on our experiences of developing and piloting approaches that support communities to actively change their local economies, it is clear that radical change is possible and can be driven from the community level, but that it needs to be supported consistently across the development context. To provoke debate we have developed five propositions for the constituent elements of an effective approach to support communities making the transition to more sustainable lifestyles.[4]

Proposition 1: The values underpinning the low carbon, high well-being local economic model need to be clearly articulated to harness a call to action. This would entail developing a more socially equitable economic framework of governance in the UK to effectively address long-term sustainability goals. This economic framework would be used as a policy tool by local authorities, and as a tool for communities to clearly articulate the values, outcomes and indicators of progress required to re-engineer the local economy. The framework serves the purpose of making explicit and shaping cultural and social norms within the locale, with a view to strengthening the sense of agency to address climate change locally. This would be achieved by demonstrating individual and community power to decide, share and act. In particular, developing this framework would require indicators for:

- local resilience, including abilities to absorb shocks, self-organize, innovate and learn;
- a re-engineered supply infrastructure that is less energy and resource intensive, and addresses issues of 'under-consumption'. This would include transport systems, waste management, energy and food systems;
- a re-engineered physical infrastructure particularly focusing on how to enable absolute reductions in consumption, such as retrofitting the existing housing stock to reduce energy demand; and
- different measures of progress including local measures of well-being, fossil-fuel dependency, levels of waste, core economy growth and sustainable social justice.

Whilst the framework of outcomes is common, this does not mean that it becomes a blueprint for change. Each community, whether rural, inner city or coastal, will find actions that are specific to its development context to achieve the outcomes.

Proposition 2: Adaptation, and hence adaptive capacity, at the community and level of the household depends critically on the institutional and macro levels above. Combating climate change can be driven from within communities, as adaptation is more likely to occur where there are shared values and understanding at a scale that can lead to more immediately felt action. However, for that action to be effective it needs to be supported by mutually reinforcing change across the development context: that is, at the institutional (structures, regulations, standards and services operating to support that community) and macro levels (policies, plans to enable adaptation opportunities, technology availability and social policies).

In Denmark, for example, grassroots innovators played a central role in the creation of a world-leading wind energy industry (Smith, 2003). Currently, over 80 per cent of Denmark's wind turbines are owned by cooperatives, local companies or individuals (Assadourian, 2008). This was facilitated by government support through feed-in tariffs, whereby renewable energy producers obtained a set price for the electricity they provided to the grid. As a result, over 20 per cent of Denmark's energy now comes from wind. Wind energy is the most commercially competitive renewable energy source. In the UK, however, it still faces a number of barriers. This includes public opposition, embedding wind generators into electricity distribution networks and land-use planning. It is important to note that local ownership and resulting local profits were key drivers to public acceptance in Denmark (Daugaard, 1997; Poetter, 2007).

Proposition 3: Development of economic and environmental literacy tools and resources is required to support awareness, understanding and action. These need to be accessible to groups across the community, support a feeling that people can be agents for change, and identify actions that will support the new economy goals. Re-engineering a local economy cannot be achieved through isolated episodes of action occurring within select sections of society. Pro-environmental behaviour requires a society-wide response, and a process that is solution-focused, supporting diverse groups across a community (and actors from across the development context) to come together. The topics of climate change and economics abound with technical jargon which acts as a barrier to involvement, reduces individuals' feeling of competency and results in passive acceptance of the status quo.

The accessibility of a process to support change has a number of facets: taking complex issues and translating them into language and representations which are generally understood and immediately relevant; engaging groups across the community to be part of the process (individuals, businesses and institutions); and the pace with which information and knowledge are shared, allowing time for reflection by those involved.

Engaging people equally in a deliberative and decision-making process; identifying achievable milestones and actions that individuals are personally passionate about and relating this back to the goals established in the framework (proposition 1), as well as celebrating success – all support a sense that individuals can effect change. Adaptation is a dynamic process, and actions and activities to support low carbon, high well-being

Low carbon, high well-being local economic outcomes framework

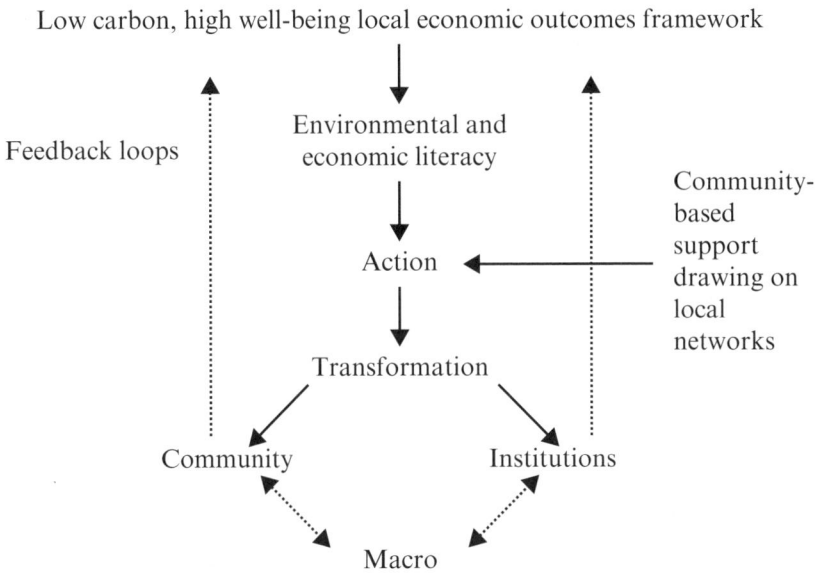

Figure 8.1 Summary of the approach to mobilizing change

economies will emerge over time. Within this process individuals, groups and institutions should be supported to experiment and innovate. This may take various forms, such as receiving approval through the decision-making process at the community level and thus sharing responsibility across a wider group; pooling resources at the community level, and mobilizing resources from institutions (see Figure 8.1). Research by Pelling et al. (2008) has shown that adaptive capacity arises out of 'social learning', which is embedded in social relationships. This is based on the theory that social ties of everyday social interaction may be a community's best resource in maintaining a capacity to change collective direction. The work provided empirical evidence to support the theoretical arguments for the contribution of relational qualities such as trust, learning and information exchange in building adaptive capacity. And, more broadly, it implies that there should be greater support to develop the core economy more generally.

Proposition 4: Motivators for action need to be opportunity-focused. Ultimately to support the level of change required needs a change in societal values, operating principles of businesses and a shift to pro-environmental behaviour. Individuals across society will be motivated to action by different imperatives. Adger et al. (2008) identify risk and perception of risk as a constraint to collective action. Our experience suggests that identifying opportunity and aligning that to a personal passion can support individuals and groups to move to action; for example, exploring potential opportunities for enterprise and jobs locally through re-engineering energy, transport and housing infrastructure.

Proposition 5: Governance mechanisms should be based within communities, have deliberative platforms for adaptive action involving wide sets of stakeholders, and equality in terms of access and control by members. To support systemic change governance structures should: embody processes of open debate to support adaptive management and learning from action; have equality of individuals as part of their operational model, and avoid institutional rigidities and vested power groups capturing value and control of key resources needed for adaptation.

Current organizational forms that could exhibit these capacities we believe are those built on mutual principles of open membership, democratic member control, co-production and equal share of benefits, and continual member learning.

NOTES

1. We note within income groups there is a high level of variance.
2. Gross Domestic Product is defined as the total money value of all final goods and services produced in an economy over a one year period.
3. The Local Alchemy programme was piloted from 2003–2007 in the East Midlands in 13 communities in rural, coastal and urban areas with populations ranging from 3000 to 28,000. See also, www.pluggingtheleaks.org.
4. Adapted from the seven principles of cooperatives; see also, www.cooperatives-uk.coop.

REFERENCES

Adger, N., N. Brooks, G. Bentham, M. Agnew and S. Eriksen (2004), *New Indicators for Vulnerability and Adaptive Capacity*, technical report 7, Norwich: Tyndall Centre for Climate Change Research.
Adger, N., J. Paavola, S. Huq and M. Mace (2006), *Fairness in Adaptation to Climate Change*, Cambridge, MA: MIT Press.
Adger, N., S. Dessai, M. Goulden, M. Hulme, I. Lorenzoni, D. Nelson, L. Naess, J. Wolf and A. Wreford (2008), 'Are there social limits to adaptation to climate change?', *Climatic Change*, **39** (3–4), 335–54.
Anderson, K. and A. Bows (2008), 'Reframing the climate change challenge in light of post-2000 emission trends', *Philosophical Transactions of the Royal Society A*, **366**, 3863–82.
Assadourian, E. (2008), 'Engaging communities for a sustainable world', in *World*

Watch Institute State of the World 2008: Innovation for a sustainable economy, New York: W.W. Norton, pp. 151–65.

Bartley, M. (2008), *Capability and Resilience: Beating the Odds*, London: ESRC.

Cahn, E.S. (2000), *No More Throw Away People: The Co-production Imperative*, Washington, DC: Essential Books.

Centre for Sustainable Energy and Community Development Xchange (2007), *Mobilising Individual Behavioural Change through Community Initiatives: Lessons for Climate Change*, report for Defra, CLG, DTI, DfT and HMT, London: Defra.

Conaty, P., E. Cox, V. Johnson and J. Ryan-Collins (forthcoming), *Decarbonising Local Economies: Local Energy Action*, London: nef.

Coote, A. and J. Franklin (2009), *Green Well Fair: Three Economies for Social Justice*, London: nef.

Cox, E. (2008), 'Think piece for the Commission for Rural Communities: economic well-being', accessed at www.ruralcommunities.gov.uk/publications/economicwellbeingroundtablenefpresentation.

Daugaard, N. (1997), *Acceptability Study for the Use of Wind Power in Denmark*, Copenhagen: Energy Centre Denmark.

Diener, E. and M. Seligman (2004), 'Beyond money: towards an economy of well-being', *Psychological Science in the Public Interest*, **5** (1), 1–31.

Dobson, A. (2000), *Green Political Thought*, London: Routledge.

Dobson, A. (2007), 'A politics of global warming: the social-science resource', accessed at www.opendemocracy.net/globalizationclimate_change_debate/politics_4486.jsp.

Edwards, R., J. Franklin and J. Holland (2003), *Families and Social Capital: Exploring the Issues*, Families and Social Capital ESRC Research Group working paper no. 1, London: South Bank University.

FAO (2001), *SEAGA Field Handbook*, accessed at www.fao.org/sd/seaga/downloads/En/FieldEn.pdf.

Green New Deal Group (2008), *A Green New Deal: Joined Up Policies to Solve the Triple Crunch of the Credit Crisis, Climate Change and High Oil Prices*, London: nef.

Hughes, A.A. and P.C. Kendall (2008), 'Effect of a positive emotional state on interpretation bias for threat in children with anxiety disorders', *Emotion*, **8**, 414–18.

Jackson, T. (2005), *Motivating Sustainable Consumption: A Review of the Evidence on Consumer Behaviour and Behavioural Change*, report to the Sustainable Development Research Network, London: Policy Studies Institute.

Johnson, V. and A. Simms (2008), *100 Months: Technical Note*, London: nef.

Kasser, T. (2002), *The High Price of Materialism*, Cambridge, MA: MIT Press.

Lenton, T., H. Held, E. Krieglar, J. Hall, W. Lucht, S. Rahmstorf and H.J. Schellnhuber (2008), 'Tipping elements in the Earth's climate system', *Proceedings of the National Academy of Sciences*, **105** (6), 1786–93.

Marks, N., S. Abdallah, A. Simms and S. Thompson (2006), *The (un) Happy Planet Index: An Index of Human Well-being and Environmental Impact*, London: nef.

Michaelson, J., S. Abdallah, N. Steuer, S. Thompson and N. Marks (2009), *National Accounts of Well-being: Bringing Real Wealth onto the Balance Sheet*, London: nef.

ODPM (Office of the Deputy Prime Minister) (2001), *English House Conditions Survey*, London: ODPM.

Pelling, M., C. High, J. Dearing and D. Smith (2008), 'Shadow spaces for social learning: a relational understanding of adaptive capacity to climate change within organizations', *Environment and Planning A*, **40** (4), 867–84.

Pielke Jr, R. (2007), 'Future economic damage from tropical cyclones: sensitivities to societal and climate changes', *Philosophical Transactions of the Royal Society A*, **360**, 1705–19.

Pielke Jr, R. and D. Sarewitz (2005), 'Bringing society back into the climate debate', *Population and Environment*, **26** (3), 255–68.

Poetter, B. (2007), 'People power, Danish style', *OnEarth* (summer edn), accessed at www.onearth.org/article/people-power-danish-style.

Raupach, M., G. Marland, P. Ciais, C. Le Quere, J. Canadell, G. Klepper and C. Field (2007), 'Global and regional drivers of accelerating CO_2 emissions', *Proceedings of the National Academy of Sciences*, **104**, 10288–93.

Reschly, A.L., E.S. Huebner, J.J. Appleton and S. Antaramian (2008), 'Engagement as flourishing: the contribution of positive emotions and coping to adolescents' engagement at school and with learning', *Psychology in Schools*, **45**, 419–31.

Rial, J., R. Pielke Sr, M. Beniston, M. Claussen, J. Canadell, P. Cox, H. Held, N. De Noblet-Ducoudré, R. Prinn, J. Reynolds and J. Salas (2004), 'Non linearities, feedbacks and critical thresholds within the earth's climate system', *Climatic Change*, **65**, 11–38.

Roberts, S. (2008), 'Energy, equity and the future of the fuel poor', *Energy Policy*, **36**, 4471–4.

Sanne, C. (2002), 'Willing consumers – or locked-in? Policies for a sustainable consumption', *Ecological Economics*, **42**, 273–87.

Schumacher, E.F. (1993), *Small is Beautiful – a Study of Economics as if People Mattered*, London: Vintage.

Seyfang, G. (2005), 'Shopping for sustainability: can sustainable consumption promote ecological citizenship?', *Environmental Politics*, **14** (2), 290–306.

Shannon, M., E. Suldo and S. Huebner (2006), 'Is extremely high life satisfaction during adolescence advantageous?', *Social Indicators Research*, **78**, 179.

Smit, B. and J. Wandel (2006), 'Adaptation, adaptive capacity and vulnerability', *Global Environmental Change*, **16**, 282–92.

Smith, A. (2003), 'Transforming technological regimes for sustainable development: a role for alternative technological niches?', *Science and Public Policy*, **30** (2), 127–35.

Stephens, L., J. Ryan-Collins and D. Boyle (2008), *Co-production: A Manifesto for Growing the Core Economy*, London: nef.

Unruh, G. (2000), 'Understanding carbon lock-in', *Energy Policy*, **28**, 877–30.

PART III

Models of sustainable and low carbon
community activities

9. The Community Carbon Reduction Programme

Simon Gerrard

BACKGROUND

For those of us fortunate enough to live in the developed world making the transition to a lower carbon way of life requires significant reductions in greenhouse gas emissions from many aspects of present day living. Of course much of the damage has already been done and so mitigation efforts will need to be coupled with planning and adaptation for the inevitable climate change already built into the system. However, the more we can reduce our emissions the less extensive adaptation will need to be and, importantly, the less painful the impacts of climate change will be too.

This chapter focuses on one attempt – the Community Carbon Reduction Programme (CRed) – to create a subset of modern society that works together to plan and implement the transition to a lower carbon future. The main motivation driving the development of CRed was the emergence of consensus in climate science about the scale of the challenge ahead and the need for action to demonstrate how organizations and individuals might move to a lower carbon future. The UK Government through its consultations on the 2007 Energy White Paper called for unprecedented levels of partnership, though there was relatively little insight at that time as to how those partnerships should be developed. Clearly mobilizing people to change their values, attitudes and behaviours at the scale required is a huge challenge requiring combinations of carrots and sticks at both (inter)national and local levels. Yet despite the scale of the challenge CRed's sense of optimism was, and remains, strong. It was instilled at an early stage largely by the fact that the programme was financed in its early years by the East of England Development Agency (EEDA). Focusing from day one on the economic development potential of the emerging low carbon economy meant that CRed was more about innovation and opportunity than crisis and disaster.

The chapter starts by describing the evolution of CRed, its ethos and how it works. Long-term engagement and monitoring are both vital features if we are to fully understand the contribution of community-scale actions to mitigate climate change. Initially the intention was to develop a community close to the University of East Anglia and Norwich. Since then things have spread considerably. The main focus for community action is at local authority level, most notably in the East of England, the North West and Birmingham, with the expectation of more to follow. These local authorities are using the CRed System as a coordinating mechanism for their climate change strategies and plans. Additionally an office-based trial is underway to establish the value of this approach within a workplace community. If successful one might expect a sharp growth in large corporate organizations with many staff using this kind of approach.

CRed intends to run until 2025 so keeps a keen eye on the long term. The long-term focus introduces the metaphor of the journey, now increasingly widely adopted (HM Government, 2009), and the chapter outlines the existing pathways available to modern society. Making this journey successfully will rely on innovation driving the development and widespread uptake of new low carbon technologies and services to reach the overall goal of an 80 per cent reduction in carbon dioxide emissions by 2025. However, technological innovation alone is doomed to fail unless it is combined with significant behavioural change. Business as usual is not an option for avoiding dangerous climate change.

THE EVOLUTION OF CRed

The School of Environmental Sciences is part of the University of East Anglia in Norwich, UK. Known colloquially as ENV, the School has spent the past 40 years characterizing and understanding environmental systems. Some of its work has focused on earth system sciences including the carbon cycle. Other aspects have focused on the social science aspects of sustainability. Together the combination is very powerful. As environmental issues have crept up the agendas of international and national governments, multinational businesses and other important organizations, so has the need to be able to explain not just why things are happening but to predict what is going to happen and, importantly, offer policy-relevant solutions to key decision makers.

Largely inspired by the frustration of climate scientists in ENV that action to tackle the prospect of dangerous climate change was insufficient at all levels, in the Summer of 2002 the concept of CRed was shaped with the aim of creating a microcosm of modern society, including individuals, households, large and small public and private sector organizations who would work together as a community to plan and act to reduce their collective carbon dioxide emissions. By 2025 this community would have demonstrated to the rest of the world how the transition to a lower carbon future (then a 60 per cent reduction by 2025 based on 1990 levels) could be achieved.

A critical factor in its infancy was the support of the EEDA, the organization responsible for promoting economic development in the six counties that comprise the East of England region. Several natural and socio-economic features of the region combined to stimulate the excitement of EEDA's economic development officers in low carbon thinking. First there was an existing industry involved in the offshore exploitation of natural gas. Both this industry and the agricultural sector were contemplating futures with some trepidation. The region's low lying land is one of the windiest in Europe so the prospect of a low carbon economy offered some hope for future prosperity in the generation of renewable, low carbon energy; offshore wind, wave and tidal coupled with onshore biofuels. The flip side to these opportunities was the threat of increased flooding from sea-level rise. These two pressures squeezed EEDA into action culminating in a five-year funding programme for CRed on a sliding scale to zero after year five.

In November 2002 the CRed programme appointed its first staff. Over the coming months a small team of key staff, including outreach officers, data management specialists and background researchers, was tasked with creating a mechanism by which individuals, households, communities, businesses of all types and other public organizations could join up and start their journey towards a lower carbon future. This micro-community would be encouraged, monitored and progress evaluated. Given the long-term nature of CRed reducing administrative burdens was important. The mechanism developed relied on an online tool combining data collection and analysis with communication, akin to a client relationship management system. The necessity for an online approach was never intended to act as a barrier to those without internet access and many organizations which use the CRed System do so with paper versions of sign-up and pledging which can be inputted into the system at a later date. The CRed System was designed to act as an umbrella under which many different activities and actions (events, case studies, demonstrations, etc.) can be coordinated and impact recorded and followed up.

Six months later, in May 2003, CRed was launched. The launch in Westminster was intended to alert key decision makers to the existence of the programme, at that time one of the first of its kind in the UK. A second launch held locally in Norwich invited key organizations to participate.

CRed's ETHOS

The front page of the local paper on the day of the Norwich launch was an image of the Earth from space with the headline 'Together We Can'. Although the headlines of local papers are not always notable for their subtlety, this one encapsulated the ethos of the CRed very succinctly. First, that we are all in this together, each of us having some role and responsibility for contributing to the solution. Secondly, that the solution will come from myriad efforts, often small in nature, that when combined will achieve the deep cuts in emissions required. Thirdly, that the emphasis is placed firmly at the local level working at the individual, household and community scale. Finally, that this is a step-by-step journey that we can make together.

The idea of a journey or a transition has become a powerful metaphor, encapsulated in Figure 9.1, that applies to households as much as to organizations. It begins with commitment – the real commitment to change founded on the principle that things will never be the same again. Unsurprisingly many organizations and individuals still struggle with this first step. However, once the commitment is made work can begin on carbon footprinting. Clearly carbon management is not the only aspect that requires attention if a more sustainable future is to be secured. There

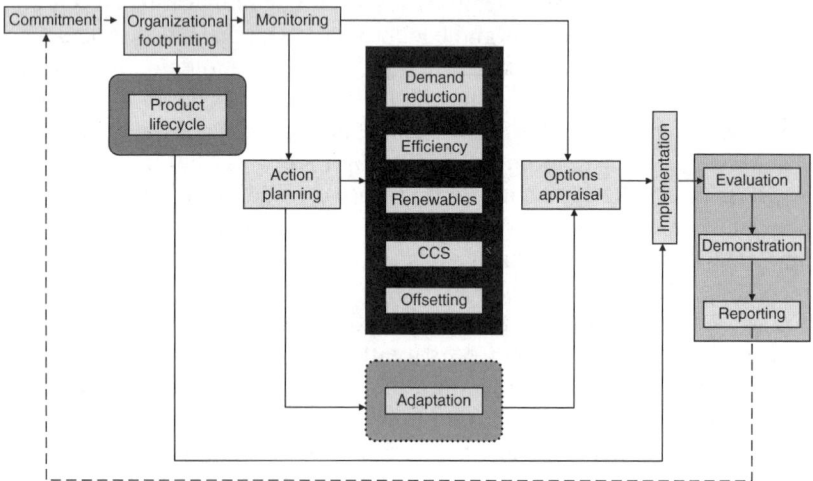

Notes:
CCS = carbon capture and storage.
© Low Carbon Innovation Centre

Figure 9.1   The carbon management process

are likely to be situations where an over-adherence to carbon management will have detrimental impacts elsewhere, for example too much focus on reducing air miles may have an impact on Fair Trade products. Similarly the uncontrolled race for biofuels stimulated by what appears at first sight to be sensible legislation in the EU and US has led to many undesirable impacts on deforestation and food supplies. However, with those important caveats in mind, there is no doubt that greenhouse gases do need to be managed and carbon is becoming a useful metric upon which to record and demonstrate progress.

Establishing baseline knowledge of current emissions is the necessary precursor to developing appropriate actions. Usually somewhat tedious work, carbon footprinting is arguably best left to those with an eye for detail and an investigative mind. Carbon is a new currency, a new language even, and our literacy is in its infancy.

Once the baseline, or carbon footprint, is in place it is time to reflect and take stock. Installing monitoring systems to check the progress of emissions is important, whether that be a simple meter reading log recorded manually or a smart meter device that collects data behind the scenes ready for analysis and presentation.

The carbon footprint is the basis for action. CRed was clear at the start and remains so today that action remains paramount. We have seen some organizations paralysed by data analysis unable to move forward until the last gramme of carbon has been accounted for. Conversely we have also seen those who jump headfirst into actions with insufficient understanding of the real problem at hand. Clearly there is a delicate balance to be struck. The journey to a lower carbon future is likely to consist of much iteration. Effective monitoring provides feedback to support this process.

THE PATHWAYS TO A LOWER CARBON FUTURE

From the outset it was clear that even if people and organizations did all that they could with current technologies and services, achieving the deep cuts in greenhouse gases required would be impossible. There was going to be a need for new technologies and services to support the journey. But getting started was important as well. CRed evolved its approach over a number of months and has developed and refined a pathways approach. The pathways describe different approaches in terms of action and effect that can be adopted to realize cuts in carbon emissions. While some are more technologically focused, others are structured around behavioural shifts and changes to institutional arrangements. In priority order the pathways are described in Table 9.1.

Table 9.1 Steps to a lower carbon future

	Description	Potential CO_2 savings
Immediate demand reduction	As it suggests, not wasting energy and emitting carbon needlessly. This primarily involves behavioural change.	10–15 per cent reduction is possible.
Improving energy efficiency	Mainly a procurement issue. Committing to buying the most energy-efficient appliances (lighting, heating, transport, white goods, office equipment, etc.) will use less energy, reduce demand further still, and emit less carbon from the use of electrical equipment and motor vehicles.	10–15 per cent reduction is possible.
Renewables	Switching to lower carbon forms of power such as green electricity tariffs can have a big impact on a carbon footprint. The challenge is to find a truly green tariff as there are many in the marketplace that are less green than they appear. There are some interesting developments in biogas from the anaerobic digestion of biodegradable wastes such as domestic kitchen waste. This could be cleaned up and fed into the existing gas grid to reduce the proportion of fossil fuel used to heat buildings and cook. Biofuels for transport have been challenging. In some instances the first generation have caused more problems than they solved. Biodiesel from waste vegetable oil is probably the most sustainable at the moment. There may be some improvements in second generation biofuels which will be designed specifically for biofuel production. However, given concerns about land required for food production it seems that biofuels for transport will at best be a partial solution.	Anywhere between 10 and 90 per cent depending on the solution in question.

Table 9.1 (continued)

	Description	Potential CO_2 savings
Carbon capture and storage	Large-scale engineering solution for trapping emissions from power stations and returning them to underground storage in existing depleted oil and gas reservoirs. Largely unproven and would require major investment in new infrastructure. One early stage development is the deliberate production of biochar, a by-product of the gasification of biomass. If returned to soil this may improve soil fertility and trap carbon, but it's still early days. The gas produced can be used to generate renewable electricity and provide heat and cooling through combined heat and power plants.	Unknown but like to be in the 30–50 per cent range.
Offsetting	Controversial, but part of a carbon reduction portfolio. This involves supporting carbon reduction activities elsewhere and accepting the emissions reductions as your own. Some programmes are better validated than others.	10–20 per cent but should be used as a last resort.

It might appear surprising to see the development of renewable energy somewhat down the pecking order in Table 9.1. That should not suggest that decarbonizing our energy supply isn't important – it clearly is. However, at the current pace of development we will fail to deliver sufficient renewable energy to meet demand. Indeed it seems unlikely that without a significant reduction in demand for energy, renewables will only ever be a partial solution to our needs – hence the importance of demand reduction and energy efficiency measures.

Initially the pathways presented organizations and individuals with a broad set of opportunities. Each pathway tends to comprise pledges that start with simple actions and become relatively more complex as the pathway evolves. For example one might start a commuting pathway by replacing the car with another option (walking, cycling, public transport) one day a week. This may extend to more days per week as familiarity and perceived benefits increase, for example feeling fitter. When the time is right the car may become obsolete and replaced by a car share scheme. Or the car may be changed for a more efficient type. The latter changes are more complex and challenging. The CRed System enables communication with people making pledges such that they can be encouraged and incentivized to make progress along the pathways.

The pathways are presented in a simple and logical order although in reality, depending on circumstances, elements of the pathways interact in potentially complex ways. For example installing solar thermal panels for hot water provision precludes the installation of micro combined heat and power. Although this is a limitation the CRed System can flag up these complexities, forewarning the user about the potential implications of their choices. Each pledge is supported by detailed information, some of which can be edited at the community level, for example directing people to specific information about public transport or grant schemes for loft insulation.

HOW CRed WORKS

At the outset CRed aimed to create a subset of modern society that would be able to demonstrate how to achieve a 60 per cent reduction in carbon dioxide emissions by 2025 – two and a half decades ahead of the then UK target. The nature and range of this subset were left necessarily vague given the open invitation for anyone to join.

Looking back over the first six years of CRed it is tempting to apply substantial levels of post-hoc rationalization to explain deliberately how progress was achieved. In reality though we were feeling our way as much as anyone else. The first couple of years focused largely on raising awareness about climate change, global warming, carbon reduction and the need for action. Though CRed is based in a pre-eminent environmental science research establishment the reliance on academic science was toned down and replaced by a more generic call for action.

However, the purpose of CRed was not simply to raise awareness but to achieve as yet unprecedented levels of carbon reduction and, importantly, to demonstrate the effectiveness of different approaches to carbon reduction. For that reason CRed developed a communications, monitoring and reporting tool enabling individuals and organizations to make commitments to carbon reduction. The combination of communication with monitoring and reporting carbon reduction was, at the time, unique. Version 1 of the CRed System was being used at the time of writing by a variety of local authorities in the UK and overseas. Version 2 was to be launched in autumn 2009 to focus more on reporting progress.

A key feature of the system is the way that commitments (or pledges) are presented to people to encourage action. Each pledge has a carbon reduction attributed to it. Once people make pledges they receive certificates reminding them of their pledge and gentle reminders encouraging them to report their success. The communication tools enable greater levels of ongoing encouragement and/or gentle nagging. Once people have completed their initial set of pledges they can be persuaded to attempt further pledges, perhaps more complex and often involving greater levels of commitment and resource.

The pledges are organized into CRed pathways along the lines of those outlined earlier (see Table 9.1). The pathways cover most aspects of daily life such as heating, lighting, travel and so on. Most people start with the easier, low or no-cost pledges and then progress at different speeds along the pathways. As people in a community, which could be residents in a local authority or staff in an organization, make their carbon reduction commitments the system tracks the levels of engagement and the potential carbon saved. Having registered with CRed, participants can access information about the pledges they have made and the carbon savings accrued. They can see their own individual impacts and that of the community in which they live or work. Local CRed operators (local authority or workplace) can communicate with individual participants, with subsets that have made similar pledges or live in the same street, or with the whole community of participants.

The potential carbon savings are just that – potential. However, once a few thousand people had joined it was possible to undertake evaluation to establish a more accurate picture of the reality of the savings. By emphasizing the importance of understanding why people hadn't been able to complete their pledges CRed was able to generate success factors for different types of pledges. These were still self-reported, but it represented a more realistic outcome than simply assuming that everyone was successful with every pledge they made. Over time repeat evaluations enable the refinement of success factors.

The self-reported nature of CRed has drawn criticism from some quarters. Of course it would be better to have absolute and accurate emissions data for every individual but this is unrealistic as the detail required to generate this on an individual basis would be prohibitively complex, requiring huge amounts of detailed data to be inputted by the participant. A balance has to be drawn between securing appropriate levels of engagement and providing sufficiently robust analysis. The CRed System fully acknowledges this and is transparent about the calculations behind the carbon savings. This enables people to query our methods and engage in dialogue about the calculations. Depending on the outcome of such dialogue the

CRed System is then updated. In the six years of operation there has been only a handful of interactions of this nature.

In reality there is not that much difference between the assumptions that lie behind the CRed System and those that have to be made about initiatives that focus, say, on the installation of loft insulation. If a local authority grant-aids loft insulation and then claims some carbon reduction there is no guarantee that the carbon reductions have actually been achieved. There is plenty of evidence to suggest that people may just prefer to live in a warmer home, adjusting to a higher thermal comfort (see for example Boardman et al., 2005).

These kinds of 'rebound effects' are well reported and present a significant challenge to those attempting to encourage carbon reduction through value shift, attitude modification and behaviour change (Sorrell, 2007). For some individuals reducing emissions in one area of lifestyle as a justification for increasing them elsewhere is not an uncommon practice. CRed's view is that rebound effects are best countered through communication focused on attempting to avoid rebound through encouragement, persuasion and support. Success will depend on changing often ingrained habits. Reinforcing the scale of the challenge while celebrating progress is particularly important. Communication is key; the CRed System is designed to enable communication at the individual level, at the group level or at the whole community level. Few organizations engaging in climate change appear to be very sophisticated in communication though there are signs that things are changing. Market segmentation approaches are becoming more commonly used and reports on rules-based approaches for communicating about climate change are being published more frequently. For example the UK Energy Saving Trust has developed an energy-saving household-based market segmentation using Experian's Mosaic segments (Business Council for Sustainable Energy UK, 2007).

A decision was made at the outset to create the CRed System as an online software tool. The prospect of tracking and monitoring thousands of carbon reduction actions across whole communities using a paper-based system would have sent shock-waves through our local woodlands. Adopting a software-based approach was more flexible, allowing for necessary evolution, and reduced considerably the administrative burden, meaning that proportionately more time could be spent on the actual engagement process. That said, many of the initial forays into the community used paper-based pledge cards simply because of the significant challenge of operating computers online but off-site. Handing out pledge cards at a local agricultural show was more practical and effective – and remains so even now.

CARBON REDUCTION IN ORGANIZATIONS

The ability for individual engagement to help an organization reduce its carbon emissions is limited. Behavioural change initiatives in the workplace can reduce an organization's carbon emissions by 10–15 per cent through things like switch-off initiatives and the use of video conferencing where appropriate (Siero et al., 1996). If commuting is included in the organizational carbon footprint then the total amount of carbon saved through behavioural change will increase as transport measures tend to be behavioural, at least to start with. However, the proportion of carbon reduction through behavioural change may well remain roughly the same as the overall carbon footprint of the organization will have increased through the inclusion of commuting.

Despite the relatively small-scale reductions a behavioural change programme is a good starting point for a number of reasons. First, it offers low and no-cost 'quick wins' that save the organization money through reduced energy use. Secondly, it generates interest in and support for subsequent carbon reduction actions that might be introduced, effectively sensitizing employees to the notion of change. Thirdly, it can generate staff loyalty through providing evidence of a responsible employer. This helps reduce the costs of ongoing staff recruitment, which can be significant in a place with a high staff turnover. Finally, a behaviour change programme can generate a sense of purpose and momentum within staff, which can lead to greater productivity and a better place to work.

Getting beyond behavioural change requires commitments made by the few on behalf of the many which cannot be easily implemented in a pledge-type system. For example installing low energy lighting requires an investment decision at board level, usually based on an assessment of the pay-back period. Once agreed the commitment is implemented by the facilities management team. Carbon reduction measures of this kind do not lend themselves to pledge-type systems. Recognizing this limitation CRed has adopted an environmental management systems approach for recording carbon reduction in organizations, linked to a case study reporting mechanism. This follows a standard 'measure, plan, act, review' approach, with the measurement of the carbon footprint highlighting both the areas of concern for carbon reduction and the need for measurement systems to provide adequate data. Implementing carbon reduction actions and evaluating their impact create case studies through which others can learn.

The rationale for this approach was two-fold. First that management systems are familiar to most organizations and so a carbon-based management system would more likely be accepted as an appropriate approach.

Secondly that reporting in a case study format might encourage other organizations to follow suit, thus creating a desired snowball effect. Case studies can be generated through the CRed System, which houses a step-by-step template for case study production. Summaries of the case studies can be uploaded to and accessed directly from the website. Carbon reductions generated by the case studies are added to the overall carbon savings generated by the individual pledges and reported on the home page of the local CRed System. These are then aggregated into total savings for the network of CRed communities.

Since its inception CRed has worked with many organizations operating at the leading edge of carbon reduction. Through time a focus on innovative approaches has been adopted which has led to the development of a £3.5 million low carbon innovation investment fund called Carbon Connections.[1] The first phase of the fund ended in Summer 2009 and invested in 27 early stage projects ranging from educational toys to smart metering systems, from alternative uses for biofuel (for example, domestic heating) to alternative methods of generating biofuel (duckweed), from new forms of wind and tidal energy to innovations in carbon capture and storage. Each of these projects provides a case study from which other organizations can learn.

FROM ENGAGEMENT TO MONITORING

The CRed System is a tool that is used within a wider engagement process. It requires a modicum of administration; about half a day per week in a typical local authority setting. However, alone the CRed System will generate few carbon reduction commitments. The system needs to be embedded within a climate change communication and engagement strategy. At first relatively few local authorities were developing climate change strategies with any significant community engagement element. This is now changing. The advent of Local Area Agreements – a kind of performance-related payment scheme – means that carbon reduction targets linked to financial payments are now more common. Though many local authorities are rightly sorting out their own internal carbon management processes first, they are beginning to focus on targets for their wider community. National Indicator number 186 (NI 186) covers emissions within the local authority boundary from domestic and non-domestic properties and transport (HM Government, 2008).

Local authorities that include NI 186 in their portfolio are set specific annual community-wide carbon reduction targets to achieve. These are typically 1–3 per cent per annum for three years. The data that support the judgements about whether or not the targets have been met are derived from gas and electricity meter data and transport surveys. These data, compiled for the Department for Environment, Food and Rural Affairs (Defra), originate from the UK's submission to the Kyoto Protocol. Data compilation focuses on the needs of the Kyoto submission first and then translates those data into a format for local authorities. This process takes time; hence the data have been 18–24 months out of date by the time they reach local authorities. This process may be accelerated somewhat but it seems likely that local authorities will be receiving outdated information on their carbon emissions for some time to come.

The time-lag creates a particular challenge for local authorities which have no way of knowing what progress they are making towards meeting their community carbon emissions targets. The CRed System, or a variant of that, may be one way to help gauge progress with sufficient accuracy to provide useful feedback to the local authority about the efficacy or otherwise of its efforts to inspire and deliver carbon reductions in its territory.

Similarly, in organizations CRed Systems can help track and monitor staff actions. CRed has launched an office-based system with a major IT company which is being trialled in the UK over Summer 2009. In this case CRed worked closely with the client to develop specific sets of pledges that relate to the company's specific management plans. These are then produced as a system which can be run either internally within an organization's own IT systems or externally hosted by CRed. Each pledge is accompanied with bespoke information and links to support changes in behaviour.

In many respects monitoring the impacts in organizations is easier as they tend to be simpler to monitor. Most buildings have only one or two energy meters which makes benchmarking and ongoing monitoring relatively straightforward. Though sub-metering between, say, lighting and computers may not always provide very good granularity some overall impact can be measured. Audits of lighting and other electronic equipment give some idea of potential loads which can be compared with the nature and range of pledges made. Reviewing expenses claims offers an insight into business travel and simple audits of car parking, cycle provision or uptake of travel passes give an indication of travel impacts for commuting.

PROGRESS

At the outset CRed's intention was to create a local community of like-minded individuals and organizations that when seen together would

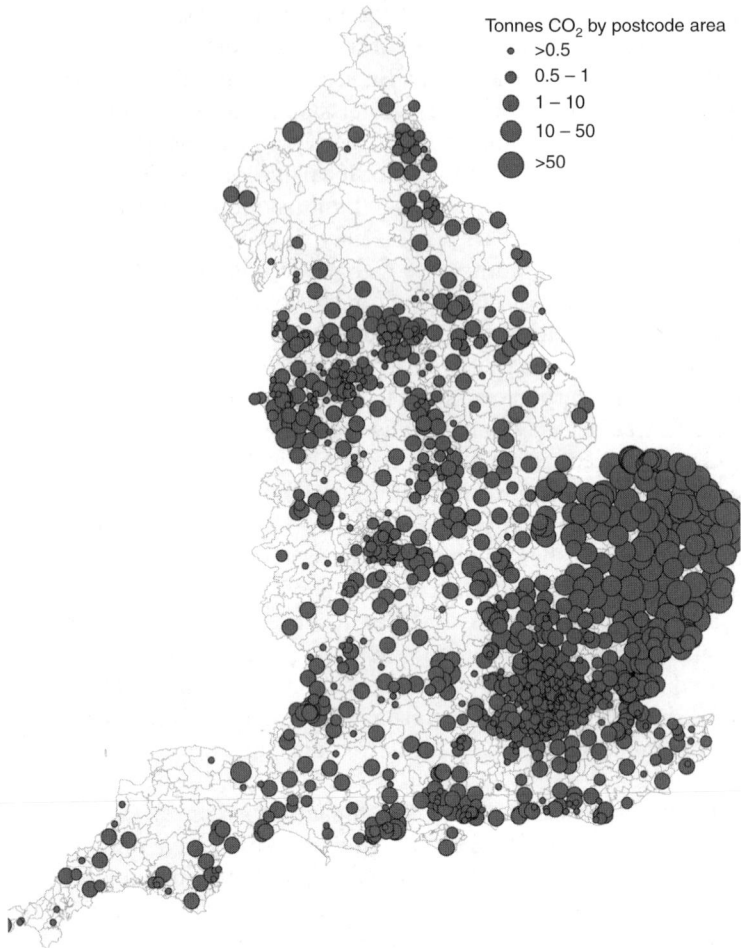

*Figure 9.2   Carbon saving through CRed engagement in England,
2003–2006*

represent a subset of modern society working together to secure a lower
carbon future. There was no expectation that the programme would
grow so rapidly. At the time of writing (the start of 2009) there are over
50,000 individuals spread across the UK that have made over 150,000
carbon reduction pledges in the network of CRed communities (Figure
9.2). There are eight local authorities running a CRed System with more
set to join later in 2009. The majority of activity is with the East and the
North West of England. CRed is now working with Birmingham and has

had a presence in London with Camden Borough Council. Interestingly, though, the map shows activity in areas where there is no local authority support for CRed, for instance in the South West. Here, I assume, people have found some value in the website alone, though that was never the intention of the programme.

Though the emphasis is on individual sign-up and action the CRed System does enable the community administrator to get an overview of community-level action. Thus one might encourage action at a street or neighbourhood level through programmes such as Global Action Plan's Ecoteams and use the CRed System to track and monitor progress. Alternatively a local authority might witness levels of individual action in a neighbourhood that prompt communication to try and join up people into a community.

Given the scale of the climate change challenge it is important that there is significant geographical spread. Although one of the pioneers, CRed certainly isn't the only initiative supporting the transition to a lower carbon world. There is a plethora of schemes at national, regional and local levels, all of which contribute to reducing carbon emissions. In some respects the myriad initiatives can lead to confusion, itself an important barrier to change. Attempts by Defra to create a UK standard carbon footprinting tool – Act On CO_2 – illustrate the recognition that we are in our infancy of carbon literacy. There is a delicate balance to tread between a doomed-to-fail 'one size fits all' top-down approach and a confusing mess of often seemingly contradictory small-scale local initiatives that start brightly and then fizzle out as funding ceases. Defra's efforts to make available the black box behind its carbon footprinting tool to others who prefer their own look and feel is very welcome indeed. It mirrors CRed's approach that each locality should expect to have its own style of communication and branding to ensure that there is a strong local element to the initiative.

CARBON REDUCTION ACHIEVEMENTS

Getting people started has been the main focus for CRed. Local authorities using the system tend to set themselves engagement targets based on new people signing up. That is a sensible approach, though after a time it becomes important to develop a parallel activity – getting those that have made their initial commitments to do more. There is an important barrier to overcome as people reach a point where they consider that they have taken sufficient action – in effect 'done their bit'. Ongoing communication and feedback on progress helps to overcome this barrier. A survey of

CRed participants in 2007 (n=844) found that 70 per cent reported that they had gone on to undertake other carbon reduction actions. Over time the average amount of CO_2 saved per person has risen from 0.8 tonnes to 1.5 tonnes. The average number of pledges has also risen, indicating that people are willing to make more commitments. However, most of the people engaged make only one set of pledges, indicating that people are happy to start but need encouragement and support to progress to other areas of their lifestyle or to make progress to more complex and challenging aspects of each pathway. This reflects the use of the system by local authorities as an engagement tool where success metrics relate to initial sign-ups. In the early stages this is fine but those that have signed up do need to be encouraged to do more.

LOCAL AND GLOBAL

The choice to run CRed as a locally driven programme stemmed from the emergence of data about low levels of public trust in central government and large businesses. That's not to say that public trust in local institutions is much higher, but there is evidence that friends, family, work colleagues, the local doctor and so on are perceived to be somehow more reliable and trustworthy.

Of course climate change is not just a local issue. Though carbon reduction may begin at home much of the embedded carbon in the products and services we use is international in scope (Carbon Trust, 2006). As this element of a typical UK resident's carbon footprint might be as high as 50–60 per cent it cannot easily be ignored. Balancing the local and global nature of the issue is important. CRed's approach is to create networks of communities that span international boundaries. To date progress has been limited to connections between like-minded universities in the United States, China, Malaysia and Japan. However, universities themselves are important players in a local community, not least through the positive economic impact they can have as an employer and home for students, but also as places of learning that tend to be respected as independent of wider political and business processes.

An international network of local communities all working together to reduce carbon emissions would be a potentially powerful instrument for engagement and action. It would help to overcome barriers about the perceived meaninglessness of action in the UK if China, India, the United States and so on are not doing their bit. It would also help to reinforce the need for action in different cultures where carbon reduction measures may be very different.

THE ROLE OF INNOVATION

Of course progress can't only be measured through geographical spread. It must also be measured by the extent of carbon reduction achieved. When CRed began we estimated that were all of the measures available to us implemented then we might achieve a 30–40 per cent reduction. Put simply, doing what we know now gets us about half the way to the target for 2050.

Inspiring, developing, manufacturing and implementing new low carbon technologies and services are vital. The exciting prospect of a 'green new deal' to help reinvent the economy and society in a lower carbon form can drive the process of change. Technologies such as smart meters can improve our energy and carbon literacy. More efficient vehicles, electrical appliances, houses and buildings will also be necessary. Increasing the pace of developing and implementing larger-scale renewable energy supplies such as offshore wind turbines, wave and tidal power, even new forms of energy crops producing sustainable biofuels, is imperative if we are to avoid the impacts of dangerous levels of climate change.

All of these innovations require the consideration of human factors. Technology does not operate in a vacuum and to be effective in application it requires careful integration with people. For example, effective smart metering requires, amongst other things, the appropriate representation of data enabling the user to understand and respond to reduce his or her energy use. Negative attitudes to the siting of renewable power plants, arguably often shaped in an insufficiently rounded single-issue framework, have hindered the ability of the planning system to deliver renewable energy supplies at a rate anything close to that which is needed to decarbonize the UK's power supply. The introduction of a hybrid petrol–electric car only reduces carbon emissions significantly if it is driven appropriately.

CONCLUSION

The transition to a lower carbon future will require significant behaviour change. This may happen primarily through the will of the individual or through the power of the community. In reality both will be necessary. Although the immediate carbon reduction achievable from behavioural change may appear to be relatively small (~15–20 per cent), it reduces the pressure somewhat on the development of renewable energy to meet demand. In meeting the challenge of climate change every little action helps.

What began as engagement has shifted to monitoring progress. The advent of carbon budgets through the Climate Change Act has introduced a new accounting focus on greenhouse gas emissions. In time the generation of carbon accounts may well lead to the introduction of personal carbon-trading systems. If so CRed-type systems that engage, monitor, feed back and report will be well placed to help evaluate the journey to a lower carbon future.

NOTE

1. See http://proxycarbon.cmp.uea.ac.uk/.

REFERENCES

Boardman, B., S. Darby, G. Killip, M. Hinnells, C.N. Jardine, J. Palmer and G. Sinden (2005), *The 40% House*, Environmental Change Institute research report 31, Chapter 4, Oxford: Oxford University Press, pp. 32–7.
BCSE (Business Council for Sustainable Energy UK) (2007), *Transforming the Market to Reduce Energy Demand*, seminar series report, London: BCSE.
Carbon Trust (2006), *The Carbon Emissions Generated In All That We Consume*, London: Carbon Trust.
HM Government (2008), *National Indicators for Local Authorities and Local Authority Partnerships: Handbook of Definitions*, revised edn, May, London: Department for Communities and Local Government.
HM Government (2009), *The UK Low Carbon Transition Plan: National Strategy for Climate and Energy*, London: Department for Energy and Climate Change, and Norwich, UK: The Stationery Office, accessed at www.decc.gov.uk/en/content/cms/publications/lc_trans_plan/lc_trans_plan.aspx.
Siero, F.W., A.B. Bakker, G.B. Dekker and M.T.C. Van Den Burg (1996), 'Changing organizational energy consumption behaviour through comparative feedback', *Journal of Environmental Psychology*, **16**, 235–46.
Sorrell, S. (2007), *The Rebound Effect: An Assessment of the Evidence for Economy-Wide Energy Savings from Improved Energy Efficiency*, Oxford: UK Energy Research Centre.

10. Global Action Plan's EcoTeams programme

Scott Davidson

INTRODUCTION

Global Action Plan's (GAP's) EcoTeams programme has been running in the UK for around 15 years. It is widely regarded as one of the most successful community based behaviour change programmes available in the UK (McKenzie-Mohr, 2009) and was recommended for expansion by the House of Commons Environmental Audit Committee (EAC, 2003). It was originally created in 1990 by a group of environmental scientists and organizational consultants who wanted to combine information, feedback and social groups as three effective behaviour change mechanisms (Geller et al., 1990).

Evidence suggests that EcoTeams may create an accelerated path of pro-environmental behaviour change for participants not only during the programme but beyond its completion too (Hobson, 2001; Staats et al., 2004; Burgess and Nye, 2008; Baxter, 2009). Some very recent data have led GAP to initiate further research into the potential for EcoTeams to motivate participants to engage in other community based environmental programmes after EcoTeams ends (GAP, 2009).

Given the positive reputation and findings surrounding EcoTeams, it is important in the context of this book and the wider literature that we offer it an in-depth examination. The aim of this chapter then, is to examine EcoTeams, its theoretical underpinnings and the evidence surrounding its impacts in more depth. In doing so, we will attempt to tease out the principles that help to make it an effective behavioural change programme, as well as the limitations of the EcoTeams approach. Using the results of this examination, the chapter concludes with a discussion of what GAP, the wider environmental behaviour change field and the national government can do to help utilize these results and overcome these limitations.

WHAT IS ECOTEAMS?

EcoTeams originated in the Netherlands in the early 1990s and since then over 150,000 people have taken part worldwide. EcoTeams has been running in the UK since the late 1990s and in its beginning involved only 200 households in Nottingham. There are now EcoTeams across the UK, and in various concentrations in Russia, North Korea, India and Lithuania as well as many other European countries.

EcoTeams in the UK works by providing training and resources for community volunteers. These resources allow volunteers (called EcoTeam leaders) to engage and motivate other people in their community in taking action to reduce their household's impact on the environment. The EcoTeam leader recruits groups of six to eight people with each person representing a different household. The group meets once a month for five months, and is provided with a set of resources to enable discussion of topics such as the environmental issues of waste, shopping, energy, transport and water. One theme each month is covered; beginning with simpler topics (for example, waste) and building up towards more challenging topics in the final months (for example, transport). With the help of these materials and their team leaders, participants discuss the issues and map out practical actions they can take to reduce their impact in each area. They are encouraged to share their experiences, local knowledge and ideas for pro-environmental action, and to support each other in making further changes to their households' behaviours. There are many subconscious and subtle social group processes going on throughout this process which help confront irrational thoughts, break habits and push pro-environmental behaviour. These processes are considered in more detail further on in the chapter when the theoretical underpinnings of EcoTeams are discussed.

Throughout the project, participants are asked to weigh their household waste and recycling output and measure their energy and water consumption. At the end of the project, these data are used to create a tailored feedback report for each EcoTeam participant. These feedback reports show seasonally adjusted energy consumption, waste, water and recycling levels over the course of the project, and outline improvements in these areas. The programme is concluded with a final EcoTeam meeting held to celebrate the collective achievements of the group, and to discuss potential next steps or future actions the participants may wish to take in their communities.

HOW EFFECTIVE IS EcoTeams?

Any discussion of the effectiveness of a behaviour change programme must include consideration of two key topics. These are impacts during and directly after the programme, and the longer term durability of these impacts. The following discussion focuses on each of these areas drawing on evidence from the relevant literature. This covers an initial summary of the two largest academic studies conducted around EcoTeams, complemented by findings from two recently completed small-scale research studies.

Evidence from the Literature: Burgess and Nye (2008)

The most recent published research investigating EcoTeams is that conducted by Burgess and Nye (2008). The focus of this research centred on 49 qualitative in-depth interviews. This was supported by 159 post participation survey responses and self reported physical data recorded across more than 1000 households as standard within the EcoTeams process. The self reported physical data included electricity and gas meter readings and the weighing of domestic waste. The research was specifically focused on drivers of actual behaviour change rather than intentions, and on durable change, that is, beyond the end of the programme (Burgess and Nye, 2008). The research will be discussed first in terms of changes made during the programme, followed by evidence around the durability of any changes made.

Qualitative evidence
In focusing on change made during EcoTeams first, interviews showed that as a result of EcoTeams participants changed their behaviour in all targeted areas. Changes were most common within the waste and water areas, with energy-related changes also mentioned frequently and transport less so. Further, changes tended to be increases of activities that were routinely being carried out already, and other behaviours which easily blended with existing lifestyles. Changes did include a smaller number of new – and some more difficult – behavioural changes however. Specific examples within domestic waste management included recycling more, considering waste when shopping (for example, avoiding excessively packaged products) and, in particular, encouraging composting. Within the area of home energy management, installing energy saving light bulbs was prominent as well as turning appliances off more and general conscientiousness regarding energy use. Some participants also reported larger investments such as installing solar water heating or better quality

Table 10.1   Frequency of mention of different behaviours during interview

Category	Action	Number
Energy	Energy saving light bulbs (or bought more)	7
	Keeping heat down/monitor electricity use	8
	Switching off lights/appliances	7
	Insulation/photovoltaics/wind turbine/gas condensing boilers etc.	5
	Switch to green energy tariff	3
	Buy efficient appliances	1
Shopping	Buy local products/lower food miles	10
	Buy organic food	3
	Grow own	1
	Reuse/reusable shopping bags	6
	Avoid excessive packaging	5
	Buy green cleaning products	2
	Buy 'fair trade' products	2
Transport	Take public transport/walk instead of car	4
	Combine journeys/share transport	4
	Avoid flights	2
Waste	Composting/green cone (or composting more)	11
	Regularly recycling (or recycling more)	10
Water	Water butts	3
	Other water saving measures	7

Source: Burgess and Nye (2008).

insulation. In relation to shopping, participants reported buying organic vegetables or local food/products, and took waste from products into account when deciding what to buy. The transport area proved the most challenging although participants did consider more efficient ways to use their cars. Some also said they were more likely to stay within the UK for holidays and use public transport if possible. Within the water theme many changes had been made. These included installing water butts and turning off taps when brushing teeth (Burgess and Nye, 2008).

Table 10.1 indicates the number of times an interviewee cited a behaviour as being adopted as a result of participating in EcoTeams (Burgess and Nye, 2008). It is likely that EcoTeam participants adopted more behaviours than they actually discussed – this is only an indication of the breadth of behaviours that EcoTeam participants adopted, rather than a comprehensive quantification.

This interview based analysis is supported both by the post participation survey data and physical data collected as standard within EcoTeams.

A note of caution is advisable here regarding the use of self reported data though. Research shows that participants can over-report behaviour/ attitudes to more closely match those attitudes/behaviours being sought by the programme. Within the environment context it is possible that participants report adopting more pro-environmental change than they have actually made. Efforts were made during this research to minimize socially desirable responding. Whilst reports may represent exaggerated responses, it is likely that they are still indicative of a high level of change due to the variety and type of data source utilized by Burgess and Nye (2008) (for example, interviews, surveys, meter readings, waste weighing and limited actual observation). This is offered some support where, on a variety of occasions during interviews, participants volunteered evidence to the interviewer of changes made, including – but not limited to – working compost bins, water butts and 'electrisave' devices. Some participants apologized for rooms being cold due to heating being turned down or offered sustainably sourced snacks and drinks to the interviewer (Burgess and Nye, 2008).

Quantitative evidence
Bearing the above in mind, results from the post participation survey conducted with past EcoTeams participants show that 94 per cent of respondents said that they were doing more now to reduce their environmental impacts than before their participation in EcoTeams. Eighty-one per cent rated EcoTeams workshops as 'effective' or 'very effective' in encouraging them to make small changes in their lifestyles. As an example of self reported actions, within the waste area 58 per cent report they now buy products that can be recycled over products that cannot and 56 per cent buy products that have minimal or no packaging over ones that do. Thirty-two per cent have started recycling phones and printer cartridges through local collection and 25 per cent compost kitchen organic waste instead of sending it to landfill (Burgess and Nye, 2008). These results are consistent with the interview data outlined previously. Further supporting data come from self reported electricity, gas and water meter readings along with waste weighing, which is undertaken as standard in EcoTeams. Figure 10.1 displays the average EcoTeams impacts calculated from these reported savings, with data collected over a three-year period.

Figure 10.1 shows that according to self reported physical data, real environmental savings are being motivated by EcoTeams. The most significant savings seem to exist within energy consumption associated with space heating and water use, with an average overall carbon reduction of 16.6 per cent per household per year. These data are in line with the mix of mainly small changes, together with some more significant changes,

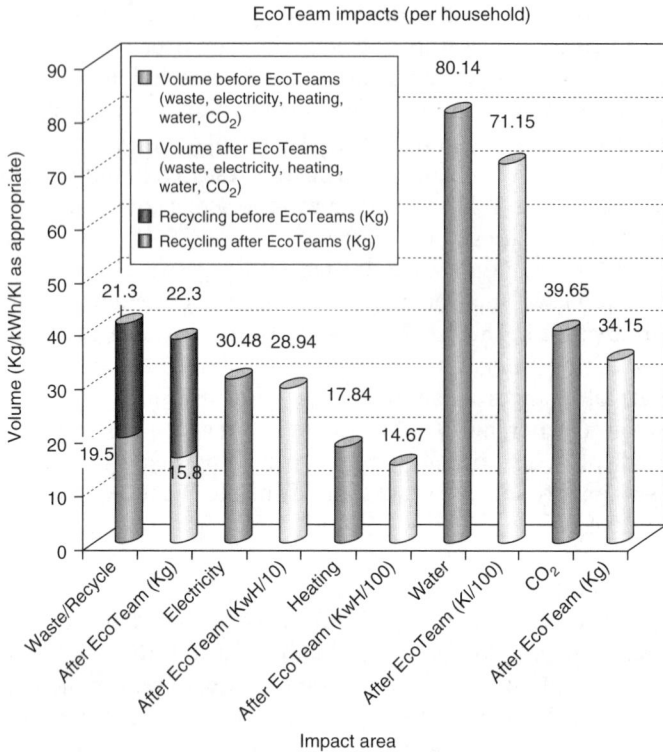

EcoTeam impacts (per household)

Source: Burgess and Nye (2008).

Figure 10.1   Average EcoTeam impacts (per household per month)

being highlighted within both the post participation surveys and in-depth interviews.

The evidence from the in-depth interviews, post participation surveys, physical meter readings and limited 'observational' data suggests that a range of simple behaviours are being motivated by EcoTeams along with some more difficult behaviours. It also suggests that these changes are contributing to significant environmental impact reductions, and that they exist across a range of areas within people's lifestyles. The following section examines the durability of these changes.

Durability of Impacts

The interviews reported above were conducted with participants who had finished EcoTeams from a period of six months previously up to three years previously. When asked if they were still engaged in behaviours they had been motivated to take up through the EcoTeams programme, all respondents reported that they had maintained the behaviours they had started, and some said they had gone on to do more. As discussed above the interviewer was also able to observe some physical evidence of changes being maintained during the interview (Burgess and Nye, 2008).

A comparison of respondents who finished EcoTeams three years ago with respondents who only finished in the last year also showed that the former were now engaged in more pro-environmental actions than the latter. Further, Burgess and Nye describe the nature of some of these further actions as 'the next level', where they require more effort and investment (Burgess and Nye, 2008). This suggests that not only are changes made during EcoTeams being maintained after it has ended, but that further significant changes are made up to three years later. More robust evidence supporting this hypothesis is discussed in the following section, which considers research carried out by Staats et al. (2004).

Evidence from the Literature: Staats et al. (2004)

Staats et al. (2004) studied 150 households participating in EcoTeams in the Netherlands. This three-year longitudinal study considered both impacts during the programme and impacts up to two years later. It selected 38 specific household behaviours to study of the 100 or so covered within the Dutch EcoTeams programme. This programme differs slightly from the UK model in that it is broken down into six themes and lasts for eight months. Another difference is that aggregated impacts of all EcoTeams in the Netherlands are included in the post EcoTeams report to participants, alongside their individual EcoTeam impacts. The range of behaviours covered in both programmes is very similar.

The researchers were particularly interested in the durability of EcoTeam effects and, as such, collected data at the beginning of the EcoTeams programme, at the end, and again at two-year and two-month intervals after it had ended. At each time point data were collected from a control group not involved in the EcoTeams programme that had been matched in pro-environmental behaviours before the programme had begun. They were matched using a subset of eight questions on domestic behaviours named the 'pro-environmental behaviour' index (PBI).

Data collection methods involved mail surveys focused on 38 specific household behaviours which were administered at all three time points. The survey included physical meter readings for electricity, gas, water and domestic waste weight measurements collected over a two-week period at each time point. It also included self report data on intentions, habits and

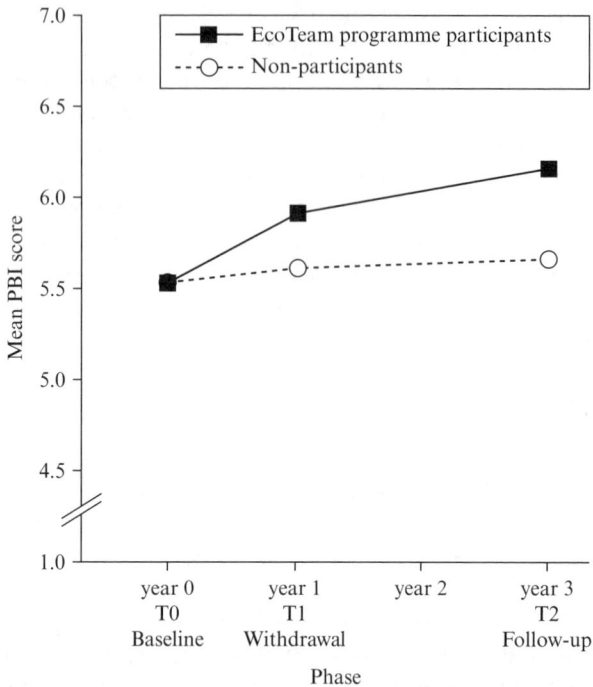

Source: Staats et al. (2004).

Figure 10.2   Pro-environmental behaviours in EcoTeam participants versus control group

social influence relating to one behaviour; using transport other than cars for journeys under five kilometres (Staats et al., 2004).

It is worth noting that the demographics of the participants are similar to those reported in the UK study, with the majority being female, university educated, middle income and aged on average from 40–55 (Staats et al., 2004; Burgess and Nye, 2008).

The self report data collected on the eight PBI household behaviour questions were used to compare levels of pro-environmental behaviour at each time point between those participating in EcoTeams and the control group. Results show EcoTeams to accelerate the uptake of pro-environmental behaviours during the programme and beyond the end of the programme up to two years and two months after ending. This is illustrated in Figure 10.2. Additionally, t-tests show EcoTeam participants to have significantly improved their pro-environmental behaviour between pre EcoTeams (T0) and post EcoTeams (T1) and post EcoTeams (T1) and two-year follow up (T2). Smaller changes within the non-EcoTeam participants were also found between T0 and T1, but not to the same extent as within EcoTeams. No significant improvements were found between T1 and T2 in the non-EcoTeams control group.

Data from the remainder of the 38 questions show improvements in 20 of 38 behaviours between T0 and T2, of which 17 are frequently performed (for example, recycling) and three are 'one time' behaviours (for example, installing an A rated appliance). Between T0 and T1, 19 behaviours changed, and between T1 and T2 11 of the earlier behaviours were maintained with 8 further advances to already changed behaviours and 1 completely new behaviour.

Supporting the self report behaviour results above, the self report meter readings and waste weighing data demonstrate significant decreases in solid waste disposal and gas consumption between T0 and T1, and decreases in all four resource consumptions measured between T0 and T2.

With regard to interaction between intention, habits and social influence on predicting behaviour, Staats et al. (2004) used hierarchical regression as their method for analysis. Results show that intention, habit and social influence did not independently predict behaviour change. Three-way interaction analyses showed that all three taken together did partially predict behaviour. The analyses suggest that where social influence was weak, intention only predicted behaviour where habits were also weak. Where social influence was strong, intention predicted behaviour regardless of the strength of habit, suggesting that one of EcoTeams' strengths may be in social influence helping to overcome habits in changed behaviour.

Before interpreting all of the above results it is important to again note the potential for socially desirable responding here. It is possible that the impacts suggested by the data above are in fact from EcoTeams increasing the likelihood for people to report that they are adopting pro-environmental behaviour and saving resources rather than from actual behaviour changes. Notwithstanding this, Staats et al. note that the results are plausible in the light of other work with similar findings around the intention–habit interaction in forming behaviour and social influence characteristic within EcoTeams (Staats et al., 2004). Further, the data across both this study and the Burgess and Nye (2008) study when combined contain similar findings across two different cultures using a variety of data sources, including quantitative, qualitative, physical and observational.

Some final data that suggests durable change as motivated by EcoTeams originate from independent studies conducted recently by Global Action Plan.

Other Studies

Recent research by Baxter (2009) involved ten in-depth interviews with
past EcoTeams participants. The interviews aimed to understand the
underlying factors which encourage or prevent individuals from progressing towards bigger pro-environmental lifestyle change, and ultimately
how GAP can assist participants in making these bigger changes. Using
EcoTeams participant survey data from Burgess and Nye (2008) and
research conducted by Defra (2008), four behaviours that have low
uptake were identified. These comprised 'avoid unnecessary flights', 'use
car less for short journeys', 'adopt a lower impact diet' and 'install micro-
generation'. Baxter conducted ten in-depth interviews exploring with past
participants each of these four themes in turn.

Findings suggest that respondents are willing to engage in more difficult
behaviours and some are already doing so. Several respondents indicated
that they had already begun actively to reduce the amount that they fly, as
a result of participating in EcoTeams (Baxter, 2009). Several also reported
starting to reduce meat consumption also as a result of EcoTeams
participation.

Predominantly motivators to the uptake of 'difficult' pro-environmental
behaviours were environmental or financial, but also included feeling a
wider sense of responsibility, or context driven motivators such as personal circumstances (for example, do not wish to fly with small children),
which affected decision-making. Barriers to the uptake of difficult pro-
environmental behaviours included a lack of coherent and consistent
information, difficulties in negotiating everyday routines in a green way,
constraints relating to symbolic consumption practices and external infrastructural constraints (Baxter, 2009).

Baxter (2009, p. 86) recommends that GAP employ a more focused
exploration of difficult behaviours within the group setting, proposing the
inclusion of a sixth EcoTeam meeting which aims to discuss the opportunities for participants to continue beyond the smaller changes discussed at
the first five meetings towards bigger, more difficult behaviour changes.

In addition to Baxter's research, initial data from recent surveys that
are core to GAP programmes suggest that EcoTeams may be motivating its participants to go on to sign up to other community based pro-
environmental projects such as 'sustainable champions' projects run
through local authorities. In a recent post participation survey around 33
per cent of EcoTeam participants surveyed reported that they had gone on
to further environmental projects and endeavours as a result of EcoTeams
(GAP, 2009). These include participation in local sustainability groups;
involvement with local school volunteering initiatives; increased recycling;

community allotment and gardening activities; and advising friends and family on environmental issues.

Data from all four studies above suggest (assuming that extremely strong levels of socially desirable responding do not exist across the survey samples) that EcoTeams motivates pro-environmental change across a wide range of behaviours, from various areas of people's lifestyles. They also suggest that not only do these changes endure, but that they lead to other more difficult changes which require additional effort and investment up to and not limited to two years after the programme ends. Recent data (GAP, 2009) also highlight the possibility of EcoTeam participants going on to be involved in other community based environmental projects which may be linked to motivation inspired by participation in EcoTeams.

We have highlighted the changes that EcoTeams is potentially motivating within its participants, and subsequently on their environmental impacts, but we have not discussed how this motivation works. There is a significant body of literature which attempts to explain why EcoTeams is motivating these changes which can form the basis of this discussion.

In both major studies discussed so far (Staats et al., 2004; Burgess and Nye, 2008) three major motivating factors were identified. Both sets of researchers cite social support and pressure, measurement and feedback, and personalized, persuasive, local information as three key underlying factors to EcoTeams' effectiveness. In addition to this, Staats et al. cite the large numbers of behaviours targeted within EcoTeams as essential in making change durable, and Hobson (2001) notes the creation of a 'discursive lens of difference' through EcoTeams which establishes an environmental aspect to thinking about daily routines and behaviours. Each of these factors is discussed in more detail below.

Social Influence

According to Burgess and Nye's (2008) analysis of EcoTeams, social support operates in the sense of reinforcing people's 'green' behaviour and reassuring people that their lifestyle choices are good through the presence of other similarly minded people. It further legitimizes behaviours through confirmatory information and other positive similar examples. Additionally, Spaargaren and van Vliet's (2000) model of pro-environmental behaviour offers support in relation to the importance of social groups. The model suggests that through the social discussion and interaction involved within groups like EcoTeams, habits can be broken as they are brought from a subconscious, automated sphere where they are not consciously considered (the practical consciousness) into an accessible area open to consideration, challenge and change (the discursive consciousness) (from Giddens, 1984; see also Phillips, 2000). Hobson (2001) argues in support of this, also citing Giddens' (1984) structuration theory, noting that participants within EcoTeams often report a 'why do I do that?' revelation after breaking daily routines down into their constituent behaviours through EcoTeams. Hobson argues that this revelation is similarly evidence of EcoTeams moving behaviours from practical consciousness into discursive consciousness. EcoTeams' potential ability to break habits through this process takes on greater importance when considered against Verplanken and Faes' (1999) outline of counter-intentional habits. Due to their automated nature (in that habits by their very nature do not receive much cognitive attention or deliberation), they are very difficult to change. They may hold an individual's daily routines unchanged, despite the individual trying to change, for example in walking to the wrong side of the car in a foreign country, despite having driven there for weeks. Of course these frustrations are all too frequent in recycling, composting, transport and other pro-environmental behaviour switches (Jackson et al., 2005), making EcoTeams' potential to break habits all the more important. If these habits are broken through the EcoTeams process, Spaargaren and van Vliet would argue that new pro-environmental habits promoted through social discussion and challenge can be formed at the discursive consciousness level before sinking back down into the automated practical consciousness resulting through repetitive behaviour. Hobson similarly asserts that a new 'discursive lens of difference' is formed through the agency of EcoTeams which allows normally automated behaviours to be constantly re-examined in light of their environmental attributes. This new durable 'lens' is potentially a key aspect in EcoTeams' impacts becoming durable, and potentially leading to other, more difficult behaviour changes.

Social pressure is suggested by Burgess and Nye (2008) to act around the basis of social identity theory (SIT) (Tajfel and Turner, 1979). SIT postulates that people who identify themselves as belonging to a particular social group will feel strong internal pressure to align their attitudes and behaviours to that group. There is a sizable literature available which argues that group social pressures can act as key driving forces in areas such as genocide, war and football-related brutality (see for example Lewin, 1947; Tajfel and Turner, 1979; Staats et al., 2004; Hunter et al., 2005). The strength of social influence is also supported by Staats et al.'s (2004) findings. They found that where social influence was strong, habit strength had no significant moderating effect on the positive relationship between intention and behaviour. Social pressure, according to a broad range of research including specifically within EcoTeams, is likely to be a key factor in motivating pro-environmental change within the programme.

Measurement and Feedback

Burgess and Nye's second suggested key factor is measurement and feedback. They cite two consequences of this. First, a consequent feeling of 'competence and control' in participants over personal environmental impacts, and secondly a 're-materialization' of the waste and energy that we produce and use (Burgess and Nye, 2008). The latter is described as the most important of the two, as it brings back into full consciousness the waste that we throw away and the energy that we waste. Without EcoTeams these concerns may not have been in participants' thoughts due to relatively automated behaviour related to them. Several examples during the interviews point to this heightened awareness, as EcoTeams participants changed what they bought as a result of considering the waste that certain purchases would produce.

The perceived 'competence and control' that the measurement and feedback provide to participants over their waste and energy behaviours are possibly key factors in overcoming barriers such as feelings of helplessness or low self efficacy. Certainly feelings of competence or 'perceived behavioural control' (Burgess and Nye, 2008) have been singled out as important in the adoption of pro-environmental behaviours (Bandura, 1977; Stern, 2000). Further, in research specifically designed to test the effects of feedback, De Young (1996) found that behaviour was more likely to last under conditions where feedback was given. De Young argues that this works through self esteem, self efficacy and reward mechanisms that suggest further pro-environmental behaviour uptake may be possible as a result of programmes which incorporate such measures.

In support of this, sports psychologists have found individual perception of how able they are to complete a physical task is a significant predictor of how successful they are likely to be in undertaking that task (Feltz et al., 2007). Measurement and feedback are likely to be further key factors in motivating change. This mechanism is likely to operate through improving self efficacy, empowerment and reintroducing waste and energy to participants' thinking.

Tailored Information

The third factor highlighted as a key functional component to EcoTeams is the local and tailored information gained through peer-to-peer group interactions, which are core to EcoTeams. This information, specific to each individual's needs, is critical in overcoming specific local barriers such as places to source environmentally friendly produce or recycle certain products (Burgess and Nye, 2008). Other research specifically focused on the importance of 'tailored' versus 'general' information supports Burgess and Nye's assertion of the importance of tailored and locally relevant information (Abrahamse et al., 2007). In examining the type and style of information given in more detail, GAP ensures information is relevant to the average householder, given in a positive style and broken down into an accessible format through a same peer level source. These factors match 'principles of persuasion' that have been highlighted as successful factors in changing behaviour in research extensively outlined by Bator and Cialdini (2000). Hobson (2001) further echoes the importance of the format of information in EcoTeams, highlighting the positive and creative use of facts around environmental impacts of behaviours. She asserts that this speaks to the participants' emergent discursive consciousness and is important in helping the participants to connect their individual experiences and subsequent environmental impacts (Bickerstaff and Walker, 2001). Tailored information, an appropriate format, delivered through in-group peers, is likely to be the third key factor in EcoTeams' potential to motivate change. It is also likely to facilitate further changes to be made that can, in combination with the group influence to help maintain changes, also add an environmental angle to peoples' decision-making process.

Interaction of Social Influence, Measurement and Feedback, and Tailored Information

Social influence, measurement and feedback, and tailored information interact to facilitate 'joined up thinking' and 'reflexive lifestyle examination', according to Burgess and Nye (2008). As such they suggest that through the sustained support and tailored information and feedback, EcoTeams facilitates a continuous re-evaluation of 'bundles of behaviours' that allows new habits to begin to be formed. This matches with Staats et al.'s (2004) suggestion that another effective aspect of EcoTeams is in the breadth of behaviours targeted, with over 100 households. If we consider behaviours not individually but as bundles we can see that interventions which only target a limited number of behaviours are potentially less likely to motivate a 'reflexive lifestyle examination' as only a few behaviours are unlikely to generalize to the whole lifestyle (Staats et al., 2004). As such, targeting small groups of behaviours is less likely to sustain change long enough to form new habits than a widely targeted programme, which might facilitate this kind of lifestyle examination. 'Reflexive lifestyle examination' is said to happen within EcoTeams through considering current everyday routines against new information coming in through peer-to-peer social groups, in a context of social pressure and support (Burgess and Nye, 2008).

In summary it seems as though EcoTeams is able to provide an environment where people receive relevant information in an appropriate format from a trusted source (peers). Additionally, this information is delivered in the context of a socially supportive environment, which also has elements of social pressure linking to social identity which helps people move automated behaviours into discursive consciousness where they are available for change and the reformation of new pro-environmental habits. Further, this new 'discursive lens' may facilitate a durable addition to thought processes where environmental consideration is given more frequently to daily routines and behaviours, allowing change to be durable and further change to happen. Finally, measurement and feedback allow barriers of self efficacy and helplessness to be overcome through empowerment, and allow key issues of waste and energy to re-enter people's discursive consciousness. The interaction of each of these key factors of EcoTeams seems to create a process of ongoing 'reflexive lifestyle examination', closely related to Hobson's concept of a 'discursive lens of difference', which encourages participants to continue considering the environment as a factor in daily routines and behaviours, and make further change beyond the end of the programme. Many of the positive aspects of the EcoTeams programme have been discussed so far in this chapter. The following section considers some of its limitations.

LIMITATIONS OF THE EcoTeams PROGRAMME

The EcoTeams programme has two major limitations:

1. The criticism that changing small behaviours will not provide the impact reductions required to avoid serious climate change impacts.
2. Problems in scaling up the programme due to cost and demand.

Considering the first limitation, it is important to bear in mind that EcoTeams was originally focused on small changes which fit into people's lifestyles and was marketed in this way, and so is open to this type of criticism. Nevertheless, the research described above suggests that the EcoTeams programmes can lead on to larger lifestyle changes such as switching to fuel efficient cars and flying less. However, research has not quantified this further lifestyle change in terms of the type of behaviour, frequency of uptake or spread beyond EcoTeams participants. As such there is little current robust evidence to adequately address this concern. Specifically related to this, the *Weathercocks and Signposts* report from the World Wildlife Fund (Crompton, 2008), and more recently *Simple and Painless* (Crompton, 2009), have given extensive attention to the problem of focusing on only small behaviours. In them Crompton argues that we cannot rely on the notion that engaging people in small changes will automatically lead to them making larger changes later. Further research is needed to quantify exactly what type of behaviours EcoTeams participants go on to carry out, and with what frequency each behaviour happens. Only when we evaluate this properly will we know the extent to which this limitation applies to EcoTeams. It is worth noting that internal research trying to quantify exactly this is currently underway at GAP. Further, GAP has recently re-written the core content of EcoTeams with a significant focus on actively moving people from smaller behaviours on to larger behaviours using an 'impact hierarchy'. Initial feedback from participants is that the hierarchy is one of the highlights of the new content and will prove extremely useful. We can only speculate however, at the moment, as to the impact it will have in motivating more significant behaviours. A two-year research project funded by Defra is currently under way which will consider the effectiveness of the hierarchy in moving people on to more significant behaviours.

Staats et al. (2004) correctly point out that due to the high demand of EcoTeams on participants and high per capita cost relative to other less intensive behaviour change programmes, scaling up and reaching beyond environmentally friendly audiences represents a substantial challenge. While Global Action Plan has reduced the demands of the programme from the Dutch model studied by Staats et al., these limitations still apply to the EcoTeams UK programme. GAP is currently attempting to address this limitation in three ways: seeking large-scale partners that can assist with the scaling up of programmes, reporting the wider benefits of GAP programmes to represent their true value against their cost, and refining the delivery model to be as efficient as possible without losing efficacy.

In seeking large-scale partners GAP has had some success in securing both Sky and EDF Energy as its charity partners alongside running international programmes with E.ON. Further charity partners that will assist with the scaling up of GAP projects are possible and continuously sought. Large-scale funding bids are also proving successful, with one such project targeting 20,000 households for EcoTeams programmes over the next two years.

On capturing the wider impacts of its programmes psychometric surveys have been developed across all of the core programmes and extensively for EcoTeams, and specific research is being commissioned on indirect beneficiaries and post EcoTeams actions. The psychometric surveys aim to capture indirect beneficiaries, community cohesion, well-being, activeness, knowledge, attitude and self efficacy impacts amongst others. Initial data from these measures are beginning to arrive and are currently under analysis. Further research into post EcoTeams behaviours is also under-way and GAP hopes that by reporting more widely on impacts EcoTeams can report a more efficient carbon saving per cost figure.

In trying to make its delivery models more efficient, Global Action Plan piloted different models to find the most effective way to engage with as many households as possible whilst yielding the best environmental results. Results from the evaluation of these different delivery models show that the 'semi-facilitated' method of engagement is beneficial in terms of cost-effectiveness, reduction in environmental impact and reach-ing a large number of households (Defra, 2008). GAP now accesses large numbers of households, in a variety of communities, through this delivery mechanism by working in partnership with local authorities or organiza-tions. Volunteer EcoTeam leaders are recruited from the organization or community and are then trained by GAP to recruit and run their own EcoTeams. Support is provided through materials, a web database and on-going advice and guidance. This model is more streamlined than the original fully facilitated model. Further, GAP has just developed its first fully online EcoTeams model. Participants can train themselves online, recruit their team online, and conduct the entire EcoTeam project online using all online resources. This has substantially reduced the cost of deliv-ering EcoTeams over the long run although it is too early to comment on its effectiveness. A thorough evaluation of this model will be undertaken over the next two years, funded by Defra. This has considerable potential to overcome the cost limitation.

Having discussed the major limitations of GAP's programmes and what GAP is doing to try to address these, it is also important to consider what the rest of the pro-environmental behaviour change field can do to address these as well as national government.

THE WIDER FIELD AND NATIONAL GOVERNMENT

Research findings around EcoTeams specifically, and pro-environmental behaviour change in general, have great potential in guiding other pro-environmental programmes within the wider field. Burgess and Nye (2008) make policy recommendations on the basis of the findings around EcoTeams described above. They recommend that pro-environmental behaviour change interventions should follow a social group based approach, using gradual changes leading into larger impact behaviours. The wider research commentary around the individual impacts of the three key factors of EcoTeams stresses the importance of the use of tailored information, delivered through in-group peers, in a social setting and in a certain format which matches the 'principles of persuasion'. It also stresses the importance of measurement and feedback, and the advantages of using all three of these in combination. There are lessons here for environmental practitioners responsible for community based initiatives. These lessons exist in both specific mechanisms that work as highlighted above, and more generally in terms of the important role that the behavioural change literature can play in the design of behavioural change programmes. All too often practitioners in the field are designing programmes based on their own understanding of what people will react to, rather than basing their design in the wealth of literature that exists on what is, and what is not, effective.

Moving on to consider the implications of these research findings for national government, policy makers have not strongly supported the community based behavioural change agenda until recently. There are several reasons as to why this may be the case although I will outline just two major reasons here. First, evidence based on community based behaviour change programmes that have actually run on a significant scale is not extensive or robust enough to warrant large-scale investment in these programmes yet. Secondly, it is possible that intensive behaviour change programmes do not offer the same coverage level and political opportunities that large-scale mass media campaigns of the same cost provide. There are two potential opportunities that may allow the community based behaviour change field to help increase the attractiveness of the community based behaviour change programmes.

First, the field must make a concerted effort as a whole to measure and report the full benefits of their programmes in a robust and valid way. While this is clearly a significant task, without accurate impact reporting, decision makers will not have the justification required to provide the policies and funding needed to choose intensive but effective behaviour change programmes before mass media campaigns.

Second, the pro-environmental behaviour change field could begin employing its behaviour change expertise to politically motivate the participants of its programmes. It is not an impossible step from asking the participants of EcoTeams to change their light bulbs and change their shopping habits to motivating them to meet with their local MPs, join political parties, learn about the different policy options or become involved in consultations. These actions are another type of behavioural change, and can slot within our current programmes as behaviours to be motivated. Global Action Plan's field is well placed to motivate this and there is increasing interest in the citizenship agenda and issues surrounding voter apathy.

In considering national government's influence, major barriers towards pro-environmental behaviour uptake are policies which do not support the changes which are being encouraged in the programmes. While EcoTeams may try to motivate the installation of micro generation, high costs along with planning permission hurdles and the lack of a feedback tariff until now have created unnecessary barriers to realizing this behaviour change. Similarly, the cheap cost of flying together with the expansion of airport capacity situated alongside ever increasing train fares make motivating a switch from flying to train travel extremely difficult for pro-environmental behaviour change programmes. National policies which encourage pro-environmental behaviour with incentives and easily implemented changes, whilst making higher impact behaviours less attractive, are an essential element to the success of behaviour change programmes.

CONCLUSION

Research suggests that Global Action Plan's EcoTeams programme creates substantial impact reductions in households through motivating a wide variety of behaviour change. Research suggests that these changes may be long lasting, and may lead on to other larger behavioural changes, whilst contributing to a wider lifestyle change. It is likely that these changes are motivated through three key mechanisms, namely social influence, measurement and feedback, and tailored information. Regardless of the suggested effectiveness of EcoTeams, the programme operates at a scale which cannot address the impact reductions required in the wider sense, suffers from concerns about addressing value based change and, as yet, is limited to the more pro-environmental section of the population.

Global Action Plan is pro-actively attempting to address the key limitations described above. Whilst it is making progress on certain issues (for example, conducting more robust research into specific areas such as post EcoTeams behaviours), challenges around the potential to upscale without compromising programme effectiveness remain. The two-year research programme funded by Defra currently underway within GAP will go some way towards examining this challenge. It will do this through evaluating the effectiveness and the durability of impacts of fully online versus face-to-face EcoTeams.

Lessons from research into EcoTeams are valuable to the wider field both in terms of specific mechanisms that may be effective in motivating pro-environmental behaviour change, and in highlighting the potential wealth of resource contained in the academic literature for practitioners. Research findings also help to remind us of the importance of government support in terms of removing external barriers through policy supportive of pro-environmental behaviour and in directly supporting community based behaviour change programmes. The wider field must improve its measurement and reporting if it is to make a stronger case to government for this direct support, and could extend its behaviour change expertise to motivating more politically orientated engagement amongst its participants. This may further strengthen the case for support for pro-environmental behaviour change programmes whilst concurrently demonstrating the impacts pro-environmental behaviour change programmes are having in motivating participants.

REFERENCES

Abrahamse, W., L. Steg, C. Vlek and T. Rothengatter (2007), 'The effect of tailored information, goal setting, and tailored feedback on household energy use, energy-related behaviors and behavioral antecedents', *Journal of Environmental Psychology*, **27**, 265–76.

Bandura, A. (1977), *Social Learning Theory*, Englewood Cliffs, NJ: Prentice Hall.

Bator, R. and R. Cialdini (2000), 'The application of persuasion theory to the development of effective pro-environmental public service announcements', *Journal of Social Issues*, **56**, 527–41.

Baxter, M. (2009), 'What are the main motivators and barriers to the uptake of difficult pro-environmental behaviours within the EcoTeams setting?', MSc thesis, The Centre for Environmental Strategy, Guildford: University of Surrey.

Bickerstaff, K. and G. Walker (2001), 'Public understandings of air pollution: the "localisation" of environmental risk', *Global Environmental Change*, **11**, 133–45.

Burgess, J. and M. Nye (2008), 'An evaluation of EcoTeams as a mechanism for promoting pro-environmental behaviour change at household and community scales', London: Global Action Plan, mimeo.

Cohen, M. and J. Murphy (eds) (2001), *Exploring Sustainable Consumption: Environmental Policy and the Social Sciences*, Amsterdam: Elsevier Science.

Crompton, T. (2008), *Weathercocks and Signposts: The Environment Movement at a Crossroads*, London: WWF UK.

Crompton, T. (2009), *Simple and Painless: The Limitations of Spillover in Environmental Campaigning*, London: WWF UK.

Defra (Department of Environment, Food and Rural Affairs) (2008), *A Framework for Pro-environmental Behaviours*, London: Defra, accessed 17 March 2009 at www.defra.gov.uk/evidence/social/behaviour/pdf/behaviours-jan08-report.pdf.

De Young, R. (1996), 'Some psychological aspects of reduced consumption behavior. The role of intrinsic motivation and competence motivation', *Environment and Behavior*, **28**, 358–409.

EAC (Environmental Audit Committee) (2003), *Learning the Sustainability Lesson*, London: EAC, accessed 29 August 2009 at www.parliament.uk/parliamentary_committees/environmental_audit_committee/eac_09_09_03.cfm.

Feltz, D., S. Short and P.J. Sullivan (2007), *Self-efficacy in Sport: Research*

Strategies for Working with Athletes, Teams and Coaches, Leeds: Human Kinetics Europe Ltd.

Geller, E.S., T.D. Berry, T.D. Ludwig, R.E. Evans, M.R. Gilmore and S.W. Clarke (1990), 'A conceptual framework for developing and evaluating behavior change interventions for injury control', *Health Education Research: Theory and Practice*, **5**, 125–37.

Giddens, A. (1984), *The Constitution of Society*, Berkeley, CA: University of California Press.

GAP (2009), personal communication, 05 November.

Hobson, K. (2001), 'Sustainable lifestyles: rethinking barriers and behaviour change', in M.J. Cohen and J. Murphy (eds), *Exploring Sustainable Consumption: Environmental Policy and the Social Sciences*, Oxford: Pergamon, pp. 191–209.

Hunter, J.A., S.L. Cox, K. O'Brien, M. Stringer, M. Boyes, M. Banks, J.G. Hayhurst and M. Crawford (2005), 'Threats to group value, domain-specific self-esteem and intergroup discrimination amongst minimal and national groups', *British Journal of Social Psychology*, **44**, 329–53.

Jackson, T. (2005), *Motivating Sustainable Consumption: A Review of Evidence on Consumer Behaviour and Behaviour Change*, report to the Sustainable Development Research Network, London: Policy Studies Institute.

Lewin, K. (1947), 'Group decision and social change', in T.M. Newcomb and E.L. Hartley (eds), *Readings in Social Psychology*, New York: Holt, pp. 330–44.

McKenzie-Mohr, D. (2009), personal communication, 11 September.

Phillips, L. (2000), 'Mediated communication and the privatization of public problems: discourse on ecological risks and political action', *European Journal of Communication*, **15** (2), 171–207.

Spaargaren, G. and B. van Vliet (2000), 'Lifestyle, consumption and the environment: the ecological modernisation of domestic consumption', *Society and Natural Resources*, **9**, 50–76.

Staats, H., P. Harland and H.A.M. Wilke (2004), 'Effecting durable change: a team approach to improve environmental behavior in the household', *Environment and Behavior*, **36** (3), 341–67.

Stern, P. (2000), 'Toward a coherent theory of environmentally significant behavior', *Journal of Social Issues*, **56** (3), 407–24.

Tajfel, H. and J.C. Turner (1979), 'An integrative theory of intergroup conflict', in W.G. Austin and S. Worchel (eds), *The Social Psychology of Intergroup Relations*, Monterey, CA: Brooks-Cole, pp. 33–47.

Verplanken, B. and S. Faes (1999), 'Good intentions, bad habits and effects of forming implementation intentions on healthy eating', *European Journal of Social Psychology*, **29**, 591–604.

11. Woking Borough Council: working towards a low carbon community

Lara Curran

INTRODUCTION

Woking Borough Council (WBC) has been recognized for its long-standing commitment to tackling climate change and protecting the environment. The Council has been awarded Beacons for Sustainable Energy (2005–2006), Promoting Sustainable Communities through the Planning Process (2007–2008) and more recently the Beacon Award for Tackling Climate Change (2008–2009).

This journey however has been long term and WBC embarked on the path to thinking globally and acting locally in the early 1990s when it adopted a new approach to energy efficiency for its own buildings. This led to substantial savings in both energy and finance and the incorporation of small scale Combined Heat and Power (CHP) units in corporate buildings in the mid to late 1990s. Energy efficiency and alternatives to conventional energy production were progressed and embedded in the Council's approach to asset and property management. To date the authority's portfolio of energy projects has come to include a range of low and zero carbon technologies, including solar photovoltaics, CHP and a demonstration fuel cell.

In 2002 however there was a shift in approach. The Council adopted a comprehensive Climate Change Strategy covering all of the services it provided. This strategy saw a shift in focus from energy saving to carbon saving and the ability to contribute savings corporately from many service areas. The adoption of the Climate Change Strategy provided fresh impetus for WBC to pursue an agenda for community engagement in these issues and 'lead by example' through the learning activities from its early energy efficiency work and experiences in its own building management.

This chapter illustrates how the Climate Change Strategy and its related projects are a central tool for communicating with the public on climate change messages and enabling the borough's citizens to engage practically in tackling climate change and positively contribute to the strategy's targets.

THINKING GLOBALLY, ACTING LOCALLY

In 2002, WBC adopted a comprehensive Climate Change Strategy integrating mitigation and adaptation. The strategy is wide-ranging in its approach, covering the whole spectrum of the borough's energy uses including power, heat, water, waste disposal, planning and transport. It was updated in June 2005 and last reviewed in 2008. The approach taken with each review has been to align reduction targets and action plans to the latest research and national policy developments. In this way, the latest strategy is aligned with the long term targets of the UK Climate Change Act of at least an 80 per cent cut in greenhouse gas emissions by 2050 against a 1990 baseline.

Considerable progress has been made since the strategy was first adopted in terms of both local activities and national policy development. The revised strategy (WBC, 2008) took account of this progress and subsequently introduced three new themes reflecting the wish to expand and strengthen WBC's approach to mitigation and adaptation; namely Water, Community and Business. The strategy covers the following ten key themes:

- Planning and regulation;
- Energy;
- Waste;
- Transport;
- Procurement;
- Education and promotion;
- Green spaces;
- Water;
- Working with business;
- Community and residents.

Consistently throughout its iterations the strategy has provided a politically and corporately supported foundation for action to address climate change locally and has always incorporated three principal objectives:

1. reduction of borough-wide CO_2 equivalent emissions;
2. adaptation to climate change;
3. promotion of sustainable development.

Due to the long term nature of the Council's engagement in the energy and sustainability agenda, it has been able to use its experiences to demonstrate leadership by example in reducing carbon emissions and developing a strategy for its own operations. In refreshing the strategy the Council introduced a theme for the community to support and empower residents and community groups to play their part in tackling climate change. The theme is informed by the views of residents and community groups. During the autumn of 2007, for example, a series of focus groups was convened and a schedule of surveys conducted to seek information on what residents currently did to tackle climate change, what they felt the barriers were to doing more and what the theme could include to help them. The objectives of the consultation were to:

- gauge current attitudes to climate change;
- gauge current levels of engagement through changes in lifestyle and behaviour;
- understand the barriers to doing more; and
- understand how best the Council could help individuals to play their part in tackling climate change.

Approximately 150 surveys were completed. When asked how Woking Borough Council could support individuals to 'do their bit' to tackle climate change, 57.9 per cent of respondents said they would like information on accredited installers and suppliers of energy efficiency measures. Respondents also stated that the Council could assist by signposting trusted information sources (see Figure 11.1). The introduction of the new theme for business recognizes the valuable contribution that local firms can make to reduce the borough's emissions. The theme was developed in liaison with Business Link and Woking Chamber of Trade and Commerce.

PARTNERSHIP WORKING

Thameswey Ltd and Thameswey Energy Ltd

Thameswey Limited is an energy and environmental services company which is wholly owned by WBC. It was established in February 1999 to promote energy efficiency, energy conservation and environmental objectives by providing energy and/or environmental services. In May 2000, Thameswey Ltd invested in its joint venture company, Thameswey Energy Ltd, to finance the first energy station in Woking town centre which officially opened in March 2001. The energy station supplies electricity by

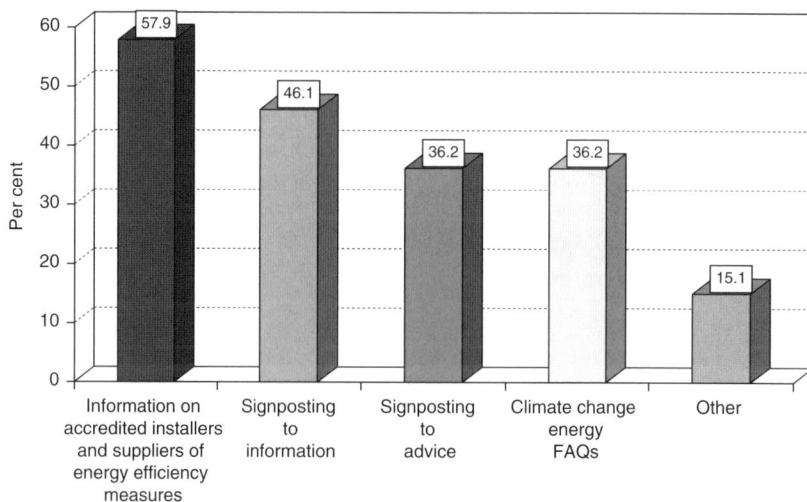

Source: Woking Borough Council, 2007.

Figure 11.1 Resident survey responses to the question: 'How could Woking Borough Council support you as an individual to do your bit to tackle climate change?'

private wire, and heat and cold water services by pipe to a number of town centre businesses and WBC's Civic Offices. In addition, electricity is supplied to other Council sites, including residential properties, by the public electricity network. Since then, Thameswey Energy Ltd has continued to deliver a range of sustainable and renewable energy projects in order to meet the Council's Climate Change Strategy objectives, which include the fuel cell CHP project in Woking Park and the combined CHP and photovoltaic system at Brockhill.

As part of the company's business plan, projects outside the Borough are also investigated. Profits can be used to improve the environment within the Borough of Woking and benefit its residents. One example is a CHP energy station for a mixed-use development currently under construction in Milton Keynes.

Climate Change Working Group

The Climate Change Strategy is overseen by a cross-party Member and officer group called the Climate Change Working Group. Its membership reflects the cross-cutting nature of the strategy. Climate change at WBC is an apolitical issue and so the Working Group has membership from both parties (Conservative and Liberal Democrat) of the Council. The business community is represented through the Woking Chamber of Commerce and residents and the voluntary community is represented through Woking Local Agenda 21 (LA21). WBC's own workforce is represented by a Member of Unison (Britain's largest public sector trade union).

The principal objective of the Working Group is to monitor and develop progress against the Climate Change Strategy's action plan and targets and to recommend decisions on new projects.

Woking LA21

At the Earth Summit in Rio de Janeiro in 1992, local authorities world-wide were challenged to address global environmental concerns by taking actions locally, under community-led Local Agenda 21 programmes. Woking Local Action 21, supported by WBC, has promoted climate change initiatives to the borough's residents since its formation in 1994 and the development of its initial action plan in 1998. Examples include:

- 2002: Woking LA21 published the Woking Green Pages guide to sustainable living which was distributed to every household in the borough.
- 2005: The group launched Greener Homes, a guide to environmentally friendly home improvements. This was made available at community venues and retail outlets across the borough.
- 2006: The Woking LA21 website was created. It is hosted on the Window on Woking community website and presents a range of projects and top tips.
- 2008: The group celebrated ten years of action at its Annual General Meeting.

As well as a key representative on the Council's Climate Change Working Group, Woking LA21 is also a linked partner of the Woking Local Strategic Partnership. This facilitates joint working with other councils, police and health providers and further promotes the integration of climate change issues in service provision.

COMMUNICATING CLIMATE CHANGE

WBC recognizes the importance of communicating the need for local action to tackle climate change. The Council references this need through a range of Climate Change Strategy actions, which aim to send clear messages to a variety of audiences including Council staff, residents, schools, developers and suppliers.

The Council aims to lead by example and inspire groups and individuals to take action to address climate change. There is a need for all Council staff to be engaged in climate change as an issue, and to be able to communicate this across the community. Several initiatives have been undertaken to encourage staff to embrace the climate change and sustainability agenda. Key examples are highlighted in the following sections.

Embedding Climate Change Action in WBC Business

The Council's Procurement Strategy encourages the purchase of products and services with reduced environmental and social impacts. An example of this is the adopted specification of environmental standards for vehicles within contracts for grounds maintenance and refuse collection. Refuse vehicles now run on Euro 4 with 5 per cent biodiesel. Project management processes have been enhanced within the Council. The process covers the full lifecycle of a project, from conception to project initiation and implementation, through to project closure. Each new project begins with the completion of a project mandate. The mandate seeks details on project objectives, timescale, risk, contribution to Council objectives (including climate change) and cost and resource implications. This helps to further embed climate change and sustainability across the Council activities and services.

Sustainability is high on the Council's agenda as a cross-cutting issue and has been integrated into corporate activity by measuring activity against a set of 16 Sustainability Themes (detailed in Table 11.1), which cover a comprehensive range of environmental, economic and social aspirations. 'Factors that contribute to climate change' is one of these themes. Guidance has been developed for staff to undertake sustainability appraisals for projects, procurement activity, performance management, committee reports and service planning.

Communicating with Residents

The Council has taken steps to communicate climate change issues to residents through a range of media and projects, often involving innovative partnership working – as demonstrated through some of the initiatives described further on in this chapter. The Council's quarterly *Woking Magazine* has been an important channel for communicating climate change issues to local residents. The magazine is distributed to

Table 11.1   Themes of a sustainable Woking

Theme number	Theme
1	Use of energy, water, minerals and materials
2	Waste generation/sustainable waste management
3	Pollution to air, land and water
4	Factors that contribute to climate change
5	Protection of and access to the natural environment
6	Travel choices that do not rely on the car
7	A strong, diverse and sustainable local economy
8	Meet local needs locally
9	Opportunities for education and information
10	Provision of appropriate and sustainable housing
11	Personal safety and reduced fear of crime
12	Equality in health and good health
13	Access to cultural and leisure facilities
14	Social inclusion/engage and consult communities
15	Equal opportunities for the whole community
16	Contribute to Woking's pride of place

every household in the borough, so is a good vehicle for wide dissemination to residents (some 39,000 households). Council officers and partners have written informative articles about a diverse set of climate change issues and solutions, including energy efficiency, light pollution and green gardening.

The environment pages on the Council's website are a comprehensive source of local information on climate change and sustainability issues for Woking residents. The area includes information about the Climate Change Strategy, Woking's green initiatives, the Council's approach to sustainability, details of key environmental services, links to Woking Local Agenda 21 and the latest environmental news. The area received over 650 unique visitors in the month of June 2009 (this refers to individual users rather than website hits).

Working With Our Neighbours

Sustainability and positive action to mitigate and adapt to climate change are priorities for many local authorities in Surrey and across the South East. Key drivers to local authority work in this area are:

- national policy requirements to address climate change, including the Climate Change Act and National Indicators;

- local requirements to address climate change, for example the Surrey Local Area Agreement and local climate change strategies.

The Surrey Improvement Partnership (SIP) is part of Improvement and Efficiency South East, which is the Regional Improvement and Efficiency Partnership, established to help the region's 74 local authorities in their drive to increase efficiency and improve services. In this way SIP is designed to bring together all of the local authorities in Surrey to deliver improvements and efficiencies in local government services throughout the county.

SIP secured funding for Tackling Climate Change Together – a project that looks to build capacity to address climate change targets and develop projects that can bring about long term improvements within Surrey and the South East. The Surrey Climate Change Partnership (SCCP) is an officer group attended by representatives of all 11 districts and borough councils and the County Council. The partnership was established to facilitate and develop joint working on climate change issues; to develop best practice, which can be shared throughout the county and region; and to create a Surrey-wide approach that can lead to improvement and efficiency.

By working together and sharing good practice, local authorities within Surrey can help to build capacity to address climate change. This will have subsequent benefits for communities within Surrey as collaborative working will further enable and empower residents to contribute to a more sustainable environment. For example, as a direct result of funding and support from SIP, the SCCP commissioned a joint Climate Change Strategy to formulate a new and far reaching agreement for Surrey local authorities. At its foundation is a core strategy which recognizes the established and ongoing work of authorities and their partners. It also aligns shared goals and objectives to better enable them to work together to tackle climate change and look ahead to 2020 and beyond.

The strategy provides a framework for: (a) establishing a consistent approach and overarching policy for the county, (b) building on work to date and planning future activity through identified work streams, (c) communicating key messages internally within local authorities and externally to residents, and (d) providing an exemplar approach for the South East region.

The SCCP believes that this approach will help build capacity and bring about efficiencies in addressing local climate change targets and help to make contributions to the regional and national picture. Regional estimates developed by Defra showed that in 2007 residents, business and visitors in Surrey produced 7330 kt CO_2. This strategy will be at the centre of reducing the county-wide environmental footprint.

MEASURING OUR PROGRESS

The local government performance framework requires local authorities to measure their progress against a series of National Indicators (NIs). Four climate change NIs have been introduced. The ways in which these are being addressed by WBC are described below:

- *NI 185*: CO_2 *reduction from local authority operations.* WBC has long been engaged in improving the energy efficiency and reducing the energy consumption of its buildings and operations. Projects have developed from small scale energy efficiency installations in the early 1990s (presence detectors for lighting, compact fluorescent light-bulbs, etc.) to the implementation of 20 sustainable and renewable energy projects (photovoltaics and CHP – see Figure 11.2).
- *NI 186: per capita reduction in CO_2 emissions in the local authority area.* It is essential that local authorities engage citizens in enabling them to play their part and help to contribute to CO_2 reductions in the local area. This chapter demonstrates projects which enable Woking residents to live a more sustainable life. The projects have been designed to enable choices and simple changes in behaviour, ultimately making it easy for people to adopt green alternatives promoting reduced resource consumption and reduced emissions.
- *NI 187: tackling fuel poverty.* The distribution of an energy efficiency questionnaire to all Woking residents in autumn 2008 enabled the Council to identify households likely to be at risk of fuel poverty and those who would also be eligible for energy efficiency grants or would benefit from future projects.
- *NI 188: planning to adapt to climate change.* The implementation of WBC's Climate Change Strategy is resulting in practical actions to reduce emissions and adapt to unavoidable climate change. Climate change is being considered in the development and implementation of all relevant strategies, plans and policies within the authority.

THE OAK TREE PROGRAMME

Woking's domestic emissions account for 41 per cent of the borough's total CO_2 emissions (DECC, 2009). Domestic water consumption in the borough equates to over 170 litres per person per day – one of the highest levels of consumption in the UK. Working with its energy company, the Energy Centre for Sustainable Communities (ecsc) Ltd, building partner

Source: Woking Borough Council.

Figure 11.2 Town centre CHP station

(Mansell Plc) and environmental partner (Woking LA21), WBC has created a programme which aims to help local homeowners improve the carbon profile of their properties. The project provides residents with an opportunity to see for themselves how different energy efficiency measures and installations work.

Oak Tree House

Oak Tree House in Knaphill (Figure 11.3) has been transformed from an ordinary three bedroom detached house into Woking's first low carbon demonstration home – a showcase for energy efficiency, renewable technology and water saving improvements to show local people the types

Source: Woking Borough Council.

Figure 11.3   Oak Tree House

of measures they can implement in their own homes to help reduce their energy use and water consumption.

The house has three different types of insulation to reduce heat loss and has a high efficiency boiler and heating controls to provide heat where and when it is needed. Much of the hot water required in the house is provided by a solar hot water system. Energy saving lights and appliances help to reduce the amount of electricity used in the house. A 2 kWp solar photovoltaic system has also been installed.

Simple water efficient shower and tap fittings reduce water consumption and a rainwater harvesting system provides water for flushing the toilet and for taps in the garden. This system and the permeable paving on the driveway also contribute to efforts to reduce the risk of flash flooding in the borough. The garden at Oak Tree House has been designed with water and energy conservation in mind, too, with drought-tolerant planting and low maintenance wildflower turf.

The refurbishment was designed to minimize the environmental impact of the house. It has been furnished using sustainable materials wherever possible and is intended to show that an energy and water efficient house can also be a welcoming, attractive and comfortable home.

Oak Tree Programme – Homes

By spring 2012 the ambition is that the Oak Tree programme will recruit 1000 households and will help each one along the pathway to becoming a 'Low Carbon Home'. It is anticipated that reductions in emissions from household energy use of 60 per cent to 80 per cent or more (equivalent to 4.0 to 5.4 tonnes or more of carbon per household per year) and reductions in water consumption of 30 per cent or more (equivalent to 50,000 litres or more per household per year) should be achievable from a range of practical and effective energy and water saving measures, as showcased in Oak Tree House, supported by simple behavioural changes.

After formal launch in late 2009, recruitment activities focused on Oak Tree House. Residents will be encouraged to implement measures within Oak Tree House in their own homes and to make further savings by using their homes more efficiently. The programme is intended to provide a pathway for all Woking residents to follow, with measures grouped in easy-to-understand packages to suit different levels of expenditure and commitment. While some residents will be cautious and start with cheap but effective and simple-to-install measures, it is expected that others who are further advanced along the 'pro-environmental' pathway will install high efficiency heating systems and micro-generation systems. Residents planning significant refurbishment projects will be encouraged to replicate the Oak Tree House approach and install measures such as rainwater harvesting and internal wall insulation. In this way the packages will be branded in themes ranging from 'acorn' to 'sapling' to 'oak tree'.

Households that have been recruited to the programme will receive basic free advice and support (provided by staff from ecsc and experts within the local community) and may be offered a 'tailored advice' service for a fee. Programme members will be provided with online tools to help them monitor their energy and water consumption and to model the potential benefits of energy and water saving measures. They will have access to a network of installers and suppliers managed by ecsc and financing to help make the more expensive measures more affordable.

Oak Tree Programme – Schools

The Oak Tree schools programme was delivered in Woking schools from autumn 2009 through a partnership between ecsc and Woking LA21, and financed by Thameswey Ltd. The programme aims to stimulate education through providing each school with an initial assessment that can be carried out by students and by signposting schools to relevant resources. Every school on the programme is entitled to a free teaching resource pack and project toolkit that focus on climate change and related issues which has been developed by ecsc in partnership with Big Foot Theatre Company.

Following initial assessment a report will be produced that provides an overview of current on-site energy consumption and makes recommendations for the installation of energy efficiency and renewable energy technologies. In addition, recommendations will be made for other measures, for example monitoring and behaviour change, to encourage more efficient use of resources such as energy and water. The report will allow each school to make an informed decision as to which technologies would be feasible for the site and the potential issues regarding their installation. Schools will also be able to take forward a more robust analysis of on-site energy use themselves.

It is hoped that the programme will not only result in positive contributions to energy use and positive environmental behaviours at school but will also enthuse children to take the messages they have learnt back to their parents and own homes, thereby encouraging wider engagement and further contributions to the borough's climate change targets.

SUSTAINABLE TRANSPORT

One of the Council's Climate Change Strategy themes is Transport. Car ownership in Woking is very high – only 15 per cent of households do not have a car, compared with 27 per cent nationally. In total there are 1.4 cars for every household in Woking, compared with 1.1 nationally. Not surprisingly the private car is the dominant means of transport in Woking. Defra's Local and Regional Carbon Estimates (DECC, 2009) showed that 21 per cent of Woking's CO_2 emissions were from road transport.

WBC Staff Transport Initiatives

The Transport theme within the Council's Climate Change Strategy has the following action: 'Revising the Council's Staff Transport Plan with a view to all Council owned vehicles, lease cars and cars used on Council business being low carbon vehicles by 2010/11' (WBC, 2008).

Since 2000, a number of initiatives have been implemented to progress this target. WBC's Transport Plan was adopted in October 2000 and has continued to develop to incorporate a range of policies to help reduce car usage and promote smarter travel choices. Initiatives include: flexible working; discounted bus travel; pool bikes and cycling allowance; and interest free loans for rail season tickets. These initiatives will help to reach the staff transport target of 10 per cent CO_2 emissions reductions by 2010/11. Monitoring of this target will be achieved through the administration of staff surveys.

In 2006, the Council's transport and fleet use was reviewed under an Energy Saving Trust fleet health check review. This report made a number of recommendations to improve the carbon footprint associated with the Council's transport usage. As a result environmental criteria were also introduced for leased cars with specific requirements on Euro and NCAP (European New Car Assessment Programme) levels and CO_2 emissions bands.

The key factors which help determine environmental thresholds to vehicles are the Euro emission rating of the vehicle and the Vehicle Certification Agency carbon-banding Point of Sale system. Before they can be sold in Europe all new car models are subject to a series of legally enforced technical tests to ensure that they meet minimum laid-down standards. These tests include standards for exhaust emissions. Emissions standards are known as Euro I (1992), Euro II (1996), Euro III (2000), Euro IV (2005) or Euro V (2008/09) (for details see EC, 2009).

Euro NCAP involves crash testing groups of similar cars to see how well the drivers and passengers would be protected in front and side collisions. The tests also assess the likely severity of injury to a pedestrian (if hit by the car). In addition to this, the Council entered into a partnership with Enterprise Rent a Car Ltd in 2007/08 to provide rental vehicles to Council staff as a replacement to the use of grey fleet vehicles.

The scheme has been successful in saving the Council money in casual user mileage payments and in terms of meeting the Council's duty of care obligations to ensure staff conduct business mileage in safe, well-maintained vehicles that meet the Climate Change Strategy and Staff Transport Plan criteria in terms of low emissions and fuel efficiency. The next stage is to extend the scheme following a traditional car club model. The WeCar initiative, which is being progressed through a partnership between Enterprise Rent A Car Ltd, Thameswey Ltd and the Council, will see the provision of a car club service operating out of Woking town centre car parks. This will enable the following benefits:

- improved access to vehicles through a central location in town centre car parks;
- telemetry installed in the vehicles to enable traditional car club membership, card access and use;

- extension of the scheme to residents and businesses;
- positive contribution to the Council's environmental and sustainability objectives in promoting alternative modes of transport;
- short term affordable access to a car to those who may not currently own one.

Car Parking

To reduce the number of vehicle movements in the town, and to encourage drivers to opt for more environmentally friendly cars and more sustainable car use, car parking initiatives have been introduced since 2006/07 in order to reduce CO_2 emissions, in line with the Council's Climate Change Strategy.

The Council has introduced differential charging levels for season tickets in the town centre car parks. The levels are based on a vehicle's CO_2 emission rating (determined by the Vehicle Certification Agency). A 50 per cent discount on the price of car park season tickets is applied for vehicles that produce the lowest emissions (CO_2 band A), and a 25 per cent discount for vehicles in band B. Vehicles with a band G rating (the highest band) pay a 25 per cent surcharge.

In addition to this an environmental levy of five pence per visit to a car park has been included in the price of a parking ticket. The additional revenue generated will be invested in carbon offset schemes and environmental initiatives in the borough as determined in the Climate Change Strategy. Two recharging points for electric vehicles have been installed in one of the town centre car parks. The points are connected to the town centre CHP energy station to facilitate the charging of electric vehicles with sustainable energy.

Taxi Licensing

For those residents and visitors that use taxis in and around the borough they can also be assured that taxis positively contribute to the sustainable transport objectives of the Climate Change Strategy by adhering to the Council's environmental criteria. From April 2008 the Council required that all (non exempt) vehicles meet or exceed the NCAP four star safety rating and Euro engine IV standard (reviewable from 2012 to continue improving standards) in order to be licensed.

It is the intention that these standards will be enhanced to Euro emissions engine standard five (Euro V) and Euro NCAP crash safety rating five stars (NCAP 5*) from early 2014. The Council will review these standards as improvements are introduced by manufacturers.

CYCLE WOKING

In June 2008, WBC and Surrey County Council were successful in securing £1.82 million from Cycling England to be spent on improving cycling facilities in the area. Woking was one of only ten towns, plus one city, to have successfully achieved the funding and the title of Cycling Town for the next three years.

The funding, together with the money that the two councils have earmarked for this project, will now be used to make progress with a range of cycling schemes. The top priorities are:

- further to improve the Woking Cycle Network to make it even more cycle friendly;
- to upgrade the Basingstoke Canal towpath to provide a high quality east–west walking and cycling route via the town centre with links to local schools and local neighbourhoods;
- to expand further activities with local schools, including more cycle clubs and cycle storage, travel plans for all schools, the universal move to Bikeability cycle training and inter school events; and
- to improve cycle storage at the rail stations.

The Cycle Woking partners intend to work closely together to increase substantially the number of cycle journeys made across the borough. The targets are as follows:

- primary schools – increase cycling from 4 per cent to 8 per cent on 2004 levels;
- secondary schools – increase cycling from 10 per cent to 15 per cent on 2004 levels;
- commuting journeys to railway stations – increase by 50 per cent on 2004 levels;
- cycle journeys – increase by 40 per cent on 2004 levels.

By raising the profile of cycling through public events Cycle Woking hopes to encourage more people on to their bikes to realize the health and environmental benefits that cycling can bring. Some examples of public events that have taken place in the first year include:

- *Cycle Woking launch*: In September 2008 the Cycle Woking partners came together to celebrate Woking's status as a cycle demonstration town. The formal launch of the plans for the forthcoming three years was complemented by a fun festival of cycling for the public, involving exhibitions, bike maintenance workshops and children's activities.

- *Cycle Stations Project*: This community-led scheme funded and supported by Cycle Woking is a social enterprise initiative which sees the transformation of old bikes to refurbished bikes by the residents of the Lakeview estate in Woking. The next phase of the project will see the introduction of a purpose built workshop to increase the number of participants and the number of bikes transformed.
- *Tour Series 2009*: Cycle Woking welcomed the organizers of the Tour of Britain to hold a high speed professional cycling event in Woking town centre on 2 June 2009. Woking was one of ten English towns and cities that were selected to stage an event as part of the Tour Series criterium races offering local people a unique opportunity to enjoy top level cycling on their doorstep captured on film by major television broadcaster ITV4. The day also saw a celebration of cycling, with music and cycling related activities, amateur races and skills sessions.
- *Get Cycling Week*: September 2009 saw the Get Cycling team visiting the secondary schools in Woking Borough to promote cycle safety and demonstrations culminating in a Saturday family roadshow in the town centre to spread the pro-cycling message. As well as the opportunity to try out conventional and fun bicycles Cycle Woking worked with its partners to provide information stands and exhibitions to enhance the cycling message.

Infrastructure and storage improvements are also planned as part of the three-year Cycle Woking programme. This includes significantly increasing the number of cycle stands in and around the borough and introducing new and improving existing cycle networks.

ADAPTING TO A CHANGING CLIMATE

While WBC's Climate Change Strategy has a series of mitigation actions looking to reduce CO_2 emissions, it is also recognized that we will need to *adapt* to the likely effects of climate change. Key findings from the UK Climate Projections (UKCP09, 2009) predict warmer summers, wetter winters and sea level rises that are greater in the south than the north of the UK. As such WBC has progressed with partners a number of practical actions to alleviate unavoidable impacts on the natural and built environment.

The Hoe Valley scheme provides a robust example of climate change adaptation. Following serious flooding around the Hoe Stream in 2000, WBC, in partnership with the Environment Agency, commissioned a detailed study of the potential for flood defences around the worst affected areas of the Hoe Stream in Westfield. In April 2007, full planning permission was granted for a mixed development of residential properties, community facilities and flood protection work centred on the former Westfield Tip. Work to progress the scheme began in July 2008.

The Council has been working closely with the Environment Agency to achieve the required level of flood protection. This will comprise a mix of earth mounds, ponds and flood walls to achieve 1 in 100-year flood protection with an additional allowance for climate change. This joint working is seen as a model approach, and has resulted in a funding agreement being signed in December 2008 which will see the Environment Agency provide £3.8 million towards the cost of the flood protection and maintain the defences for 10 years after construction.

The scheme combines flood risk management measures with added health, ecology and leisure benefits. The Hoe Stream is a site of Nature Conservation Importance, designated because of its flora and fauna and its various important habitats including wet grassland, ponds, broad-leaved secondary woodland and meadows. These form a valuable wildlife corridor that connects the south-west of the borough with the east via the urban area. The scheme will provide a publicly accessible wildlife wetlands reserve and increase the biodiversity of the area. It incorporates attractive walk and cycle ways along the Hoe Stream and opens up further woodlands in Woking Park. Almost 58 acres of green leisure space will be revitalized for public use. The development of new 'wetlands' in the Hoe Stream floodplain area will be a significant opportunity to enhance wildlife, landscape and informal recreational opportunities in the area, while also achieving restoration of the former tip land and provision of additional flood plain capacity to help alleviate the future threat of flooding in the Westfield area.

Another site incorporating sustainable development techniques is Brookwood Farm. The site comprises a small complex of farm buildings and three vacant cottages. The development is being undertaken on behalf of the Council and is for 12 semi-detached houses. Work started on site in May 2009. The houses are to be constructed to an Eco-homes 'excellent standard' using sustainable materials. All houses will be served by a local CHP generator. Wherever possible individual property and communal areas have been designed to allow rainwater to percolate the ground. Rainwater harvesting will be installed for use in the garden and one property will trial its use in flushing toilets. The development is expected to achieve 23 per cent savings when compared with properties of a similar size and scale built to conventional standards. It is intended that there will be no net increase in water discharging to the existing storm water sewer system. The excess water will be stored on site and discharged via a controlled water outlet into a herring bone system of underground drainage pipes, creating a widespread wetland area to the west of the site.

CONCLUSION

Domestic emissions account for 41 per cent of Woking's CO_2 emissions (DECC, 2009). The Council has worked hard to reduce emissions associated with its own buildings. Together with the continued help of the community and local businesses the Council can continue to make a positive difference to the borough's impact on climate change.

At the heart of engaging the community on climate change is the need to recognize the diversity within and then to offer a range of tools and opportunities to facilitate successful participation. This approach is embodied in the range of projects and support services offered by the Council and its partners from simple signposting to more information to facilitating the installation of a renewable energy technology. The effectiveness of community engagement in climate change is enhanced through partnership working and this is demonstrated strongly in the examples considered in this chapter.

Through the Climate Change Strategy, the Council and its partners will continue to work together to promote projects that secure and provide for long term reductions in the borough's environmental footprint and CO_2 emissions.

REFERENCES

DECC (Department for Energy and Climate Change) (2009), *2007 Local Authority Carbon Dioxide Emissions*, London: DECC, accessed at www.decc.gov.uk/en/content/cms/statistics/climate_change/climate_change.aspx.
European Commission (2009), *Transport and Environment: Roadvehicles*, Europa, website of the European Commission, accessed at http://ec.europa.eu/environment/air/transport/road.htm.
UKCP09 (UK Climate Change Impacts Programme) (2009), *UK Climate Projections Version 2*, Oxford: UKCP, accessed at http://ukclimateprojections.defra.gov.uk/content/view/516/500/.
WBC (Woking Borough Council) (2008), *Think Globally, Act Locally: Climate Change Strategy*, Woking, UK: WBC, accessed at www.woking.gov.uk/environment/climate/Greeninitiatives/climatechangestrategy/climatechange.pdf.

12. Intentional community carbon reduction and climate change action: from ecovillages to transition towns

Joshua Lockyer

Public awareness has risen to such an extent that climate change is not just a topic of conversation but a call to action to make major changes in consumer lifestyles.
(Crate and Nutall, 2009, p. 11)

Climate change makes this carbon reduction transition essential. Peak oil makes it inevitable. Transition initiatives make it feasible, viable and attractive.
(Transition Towns, 2009)

INTRODUCTION

The Earth's climate is changing with potentially grave consequences for human societies. Despite growing scientific consensus that the rate and nature of climate change are due at least in part to human activities, policy-driven attempts to alter these activity patterns and reduce our collective carbon footprint continue to fall far short of their goals. Climate scientists suggest that if we are to avert climate catastrophe, we must greatly reduce the rates at which we burn fossil fuels, cut down forests and disturb our soils. Yet, more than a decade after Kyoto, global carbon emissions continue to grow (Intergovernmental Panel on Climate Change, 2007). In the light of the fact that a wide variety of government-sponsored climate initiatives have failed to bring about appropriate action to slow the emission of greenhouse gases, it is useful to explore local, community-based carbon reduction initiatives and consider ways that scholars and government officials can support and build on their efforts.

Throughout the developed world where consumption patterns are a root cause of carbon dioxide emissions, groups of people are coming together in small-scale, intentional community groups to voluntarily and deliberately change their lifestyles and reduce their carbon footprints. Growing movements of ecovillages and transition town initiatives are bringing people together to cooperate, share resources, skills and knowledge, and support each other in their efforts to live in more climate-sensitive manners. Although these ecologically and climatically oriented intentional community movements are relatively new, their numbers are growing and a small number of studies suggest that they have found effective ways to take positive climate change action (Assadourian, 2008).

This chapter provides an overview of ecovillage and transition towns movements. It describes some of the specific ways in which people in these movements pursue carbon reduction, drawing out lessons for government bodies that wish to facilitate and encourage carbon reduction and climate change action in local communities. Finally, the chapter suggests that scholars and government officials should seek engaged partnerships with these community-based movements in order to conduct research aimed at measuring the effectiveness of this type of community-based climate change action and facilitating the scaling up and dissemination of such activities within existing communities.

CARBON REDUCTION AND CLIMATE CHANGE ACTION IN ECOVILLAGES AND TRANSITION TOWNS INITIATIVES

Although the advent of the term 'ecovillage' dates back to the 1970s, it did not come into widespread use until the 1990s (Bates, 2003). In 1990 at the behest of the Gaia Trust, Robert and Diane Gilman wrote a report called *Ecovillages and Sustainable Communities*. In it, they defined an ecovillage as 'a human-scale, full-featured settlement in which human activities are harmlessly integrated into the natural world in a way that is supportive of healthy human development and can be successfully continued into the indefinite future' (cited in Dawson, 2006, p. 13). The Gilman report portended a surge of intentional community building that set its sights squarely on attempts to develop sustainable, ecologically minded lifestyles. In 1994, the Global Ecovillage Network (GEN) was formed to serve as a networking body for the various ecovillage-like experiments that were coming into being around the world.

Since that time, the number of ecovillages has grown steadily and spread to six different continents (Lockyer, 2007). As of this writing, the GEN lists over 400 ecovillages in its database (Global Ecovillage Network, 2009). While this number includes projects that are in existence as well as those that are in the process of forming, it is probably still a conservative estimate. For one thing, there are a number of ecovillage projects that do not wish to be publicly identified or included in the GEN's database. Further, there are many more intentional communities that are engaged in many of the same types of activities as ecovillages but have chosen not to identify as ecovillages or not to align themselves with the GEN. For example the Fellowship for Intentional Community's database lists over 1000 intentional communities, many of which identify as being centred on environmental sustainability activities while not calling themselves ecovillages (Fellowship for Intentional Community, 2009). So what do these communities have in common?

In contrast to most developer-designed neighbourhoods and communities that tend to replicate existing modernist designs in new settings, each ecovillage is a unique creation situated in its particular socio-cultural, historical, ecological and geographical context. Ecovillages are created by self-selected groups of people who deliberately come together to create an intentional living environment centred on values of sustainability and cooperation. As such, it is difficult to accurately characterize ecovillages in any general sense. Still, ecovillages do share a number of general characteristics and most of them are engaged in similar kinds of activities.

Ecovillages are not a new phenomenon; rather they are a specific, contemporary manifestation of what both scholars and movement participants refer to as intentional communities. Communal studies historians trace the history of intentional communities back over 2000 years (Metcalf and Christian, 2003; Metcalf, 2004). While this history is not within the remit of this chapter, a few familiar examples include the Essene monastic community where some believe Jesus of Nazareth formed some of his early conceptions of Christianity; the Hutterite communities which have spread through parts of North America over the last 250 years after their exodus from Europe and Asia; the Amana and Oneida communities of early America; Robert Owen's New Lanark community in nineteenth century Scotland; Israeli kibbutzim, and the hippie communes of many parts of Europe and the US in the 1960s and 1970s.

All these groups share a number of characteristics. People who form intentional communities live together in a geographical place in order to achieve some common purpose or goal that often arises in response to critical assessments of the dominant culture or society. These groups often have a specific set of cultural, social, political, economic and/or spiritual alternatives in mind when they form their community. They are aware of themselves as a group defined by shared sets of norms, values and practices that set them apart from the mainstream society. Their communities consist of a number of unrelated individuals, among whom there is a high degree of social interaction, some amount of economic sharing, and a degree of altruism or suppression of individual choice in favour of the good of the group or, in the case of ecovillages, the wider society and even future generations (Miller, 1999; Lockyer, 2007). Scale is also a defining characteristic of intentional communities and ecovillages; most of them consist of a few dozen to a few hundred individuals sharing a circumscribed space within which they interact in order to govern themselves and ensure a high level of adherence to shared community norms. As we face the challenge of mitigating climate change, these characteristics provide a basis for creating fundamental changes in values and practices that are essential to reducing our collective carbon footprint.

Today's ecovillages share these characteristics and have in common a general concern with creating more equitable and sustainable ways of living. While many ecovillagers share a desire to live in a way that reduces fossil fuel use and greenhouse gas emissions, it should not be assumed that people form or join ecovillages simply to live in a way that reduces one's carbon footprint; explicitly addressing climate change is only one component of their broader endeavours to create holistically different lifestyles. My ethnographic research reveals that there are a wide variety of factors that motivate people to become ecovillagers. In addition to explicitly articulated critiques of the unsustainable and inequitable nature of modern, industrial society, ecovillagers are motivated by a desire to live according to their values and to live in a more intimate and supportive social community. People's explanations of their decisions to live in an ecovillage vary greatly and may include intensely spiritual or deeply personal reasons. In addition to trying to create a more sustainable lifestyle, they also seek a way of life that is more satisfying and meaningful (Kirby, 2003; Lockyer, 2007). With that in mind, most ecovillages are engaged in a number of similar types of activities that contribute to positive climate change action and carbon reduction. These will be described below with reference to Earthaven Ecovillage in the US and Findhorn Ecovillage in Scotland.

CLIMATE CHANGE ACTION AT EARTHAVEN ECOVILLAGE AND FINDHORN ECOVILLAGE

Earthaven Ecovillage was founded by 12 people in 1994 in the Appalachian Mountains of western North Carolina. Today, it is a constantly evolving experiment in sustainable living and cultural change. At Earthaven, approximately 60 members are engaged in a variety of intentional living arrangements on their collectively owned 320 acres of forested land. Findhorn Ecovillage was started by 3 people in 1962 on Scotland's northern shores near Forres. Today, it has expanded to include over 500 participants who share a variety of levels of commitment to the Findhorn Foundation's experimental and educational missions. Both Earthaven and Findhorn are dedicated to serving as experimental models for a sustainable society. While neither Earthaven nor Findhorn focus their endeavours solely or explicitly on carbon reduction, the imperative of taking action to mitigate the worst effects of climate change is part of their larger vision and is reflected in a number of characteristic community-wide initiatives.

Economic Localization

A major point upon which many ecovillagers agree is that the structure of the global, industrial economy is out of sync with and detrimental to the continued functioning of the biosphere and its myriad ecosystems. Ecovillagers recognize that for much of human history, the most enduring cultural and economic systems have functioned largely within and acted as part of local ecosystems. In contrast, global industrial capitalism's claim to have transcended ecological constraints is clearly starting to unravel, especially as the effects of climate change begin to impinge on industrial-scale, globalized agricultural production systems. As an alternative to a global industrial economy dependent upon the burning of large amounts of fossil fuels for the production, processing and transportation of agricultural and consumer goods, ecovillagers are taking steps toward economic relocalization. They are developing locally and regionally based economic systems that reduce the distance goods must travel between sites of production and consumption, in the process reducing the carbon footprints of these commodity chains. Ecovillagers' economic relocalization efforts encompass a variety of things, including local-scale, organic agriculture, the use of alternative currencies and bioregional networking.

Ecovillagers have long been at the forefront of local, organic food movements that have gained widespread popularity in recent years for their social, ecological and physical health benefits. Local, organic food systems align well with ecovillagers' attempts to live in more socially and ecologically sustainable manners; they cut out the various agricultural middlemen, place value on small farmers and polyculture production, and provide the consumer with wholesome food and direct social connections with food producers. Local, organic food systems directly address issues of carbon reduction by eliminating much of the fossil fuel consumption that is involved in applying fertilizers, herbicides and pesticides, and in processing, packaging and shipping agricultural commodities around the world.

The members of both Earthaven and Findhorn are directly involved in the development of local and regional food systems that exist within and extend beyond their individual boundaries into the surrounding communities. In 1994, Findhorn began developing a 15 acre community-supported agriculture farm that currently supplies over 70 per cent of the community's fresh produce requirements and reaches over 200 households (Findhorn Foundation, 2009). In addition, Findhorn established a partnership with a nearby organic farm that supplies most of their demand for dairy, meat and eggs. Earthaven has followed a similar path. Over the last 15 years, they have cleared almost 10 acres of land for local-scale agriculture. The trees that they felled in the process have been used to build and heat their homes. The clearings are now used as a source of livelihood for community farmers who are raising a variety of livestock, edible and medicinal plants, and biofuels stock for local consumption both within and beyond the boundaries of their community (Lockyer, 2007; Earthaven Ecovillage, 2009).

These kinds of activities are characteristic of many other ecovillages and intentional communities around the world and clearly have the potential to serve as models for carbon reduction. Of the 1026 existing intentional communities and ecovillages listed in the Fellowship for Intentional Community's online directory, 256 indicate that they grow over 50 per cent of their food (Fellowship for Intentional Community, 2009).

Ecovillages like Earthaven and Findhorn also attempt to promote economic relocalization through the development of alternative currency systems. Earthaven refer to their alternative currency as 'Leaps', which is a reference to the leap of faith that is required to make the decision to implement and participate in an alternative to the dominant dollar economy. At Earthaven, Leaps serve not only as a form of paper currency; the Leaps system is also a way of keeping track of ecovillage members' community service hours. Each member of Earthaven is required to perform 1000 'leapable' hours of community service within their first ten years of membership. This includes things like physical labour on the installation of common infrastructure and performing duties as part of the various committees that constitute Earthaven's internal governance system. Alternatively, Leaps can be purchased at the rate of $7 per hour. Once a member has performed (or purchased) the required 1000 hours of community service, they are subsequently paid in paper Leaps which can then be exchanged for goods and services provided by other community members, thus giving everyone a number of options for accumulating their required 1000 Leaps. As the number of Leaps in circulation grows, Earthaven hopes that they will provide a foundation for building a local, resilient economy that is not directly linked to or dependent upon the dollar.

Findhorn's Ekos system functions more strictly as a form of local currency. At Findhorn much as at Earthaven, Ekos are seen as a way of encouraging the patronage of local businesses in order to generate community wealth and strengthen local economic and community relationships. Ekos come in one, five and 20 denominations and can be purchased at the community's Phoenix store or at the Eko Exchange building for the equivalent in pounds sterling. Likewise, Ekos can always be redeemed at these locations for pounds sterling. While they are in circulation, Ekos can be exchanged among members and visitors for goods and services while the accumulated pounds sterling accumulate interest and can be issued as loans to community members and to the community itself when they need to conduct business in standard currency outside of their local economic network.

Alternative currencies like Earthaven's Leaps system or Findhorn's Ekos have a long history, dating back to Robert Owen's Equitable Labour Exchanges of the early twentieth century and the local scripts that came into use during the Great Depression. These scripts were utilized as means of maintaining functioning economies on a local scale and keeping families fed and employed during economic downturns. Today, they are seen as community building mechanisms that exert social and ethical regulation over economic activity (Raddon, 2003; North, 2006). In the case of ecovillages, that often means building economic structures and networks that function on a local level, are embedded in particular social relationships, and place a priority on ecologically sustainable products and services. As Earthaven explain, 'Leaps . . . are a means to facilitate the exchange of labor and goods within our community. Ideally, they will allow us to prosper by creating our own economic system, which can flourish without being dependent on the global/industrial economy' (Earthaven Ecovillage, 2005). Similarly, each Findhorn Eko note reads in part 'issued . . . to support our local economy, promote local enterprise, and publicize local initiatives'. Thus, Earthaven's Leaps and Findhorn's Ekos systems challenge the moral ideologies inscribed in the global industrial political economy. They are alternatives that put sustainable local development and ecological values ahead of an abstract financial bottom line. By encouraging the relocalization of economic activity, these systems reduce the amount of fossil fuel use and carbon emissions associated with the economic systems upon which we depend.

Clearly, it is difficult to meet people's total food requirements, much less their needs for other kinds of goods, solely through these local, community-based organic agricultural systems or through businesses that participate in alternative currencies. Unable to meet all their needs within their communities, many ecovillagers place a strong emphasis on obtaining the goods that they do need within their local regions to the extent possible, again reducing the amount of fossil fuels consumed in transporting goods around the world. For this reason among others, many ecovillagers have aligned themselves with the bioregional movement.

Bioregionalism is a diffuse movement that encourages economic and political relocalization and the development of a greater sense of place according to ecological, rather than existing political or economic boundaries. Bioregionalists are not only concerned about political and economic relationships, but about developing cultural knowledge and practices specific to local regions characterized by particular cultural histories, weather patterns, geographical features and ecological conditions. Bioregionalists emphasize the importance of 'communities of place' whose inhabitants recognize that their political and economic activities are inherently constrained by the unique cultural and ecological characteristics of their region. As such, bioregionalism presents a challenge to the neoliberal political economy characterized by fossil-fuelled global-scale flows of goods and finance (Sale, 2000; Carr, 2004).

People who adhere to this philosophy seek to develop economic relationships with a variety of local and regional producers as an alternative to a globalized economy. Further, these relationships are shaped by the opportunities and constraints afforded by local ecological systems. For example, both Earthaven and Findhorn have developed partnerships with nearby farmers, bakeries and dairies. Often, these exchange relationships are based on the seasonal availability of particular foodstuffs. Rather than seeking to create a year-round supply of the same kinds of foods shipped from around the world, bioregionalists believe that living according to the dictates of the seasons as they manifest in a particular region is a fundamental part of inhabiting a particular place.

Bioregionalists don't see seasons as an obstacle to overcome, but rather as an inherent part of the natural cycle that must be worked with. Similarly, the geographical specificities of natural resource availability are also something that bioregionalists seek to embrace. Thus, agriculturalists at Findhorn and Earthaven often attempt to cultivate and exchange resources that are naturally adapted to the particular climates of their home regions. For example, one member of Earthaven owns and operates a medicinal herbs business whose products are all based on locally available plants. An explicit effort is made to develop products made from these local resources that can substitute for products produced from resources that might only be available outside the region and thus require extensive amounts of fossil-fuelled shipping.

These bioregionally networked economic relationships simultaneously ensure ecologically minded local producers with a steady market while also enabling ecovillagers to access goods that aren't produced within the ecovillage while not depending on foodstuffs that have been shipped hundreds or thousands of miles. This philosophy presents a challenge to contemporary, industrial society and its attendant carbon footprint because so many of the things that we have grown accustomed to consuming are extracted, produced and transported on a global scale using fossil fuels. This is one of the fundamental challenges that ecovillagers and bioregionalists are attempting to address and, to the extent that they are successful, they provide models for carbon footprint reduction.

Eco-technologies: Renewable Energy and Energy Efficient Housing

As the links between fossil fuels and climate change have become clear, a growing interest in alternative, renewable energy sources has emerged. Much as ecovillagers pioneered local, organic food movements, they have also been at the forefront of experiments in renewable, sustainable energy technologies. Ecovillagers have moved away from fossil energy dependence by employing photovoltaic, wind power and biofuels technologies on a local scale. These voluntary, intentional experiments have the potential to serve as data sources and models as we seek ways to meet our energy needs in sustainable, less carbon-intensive manners.

Earthaven Ecovillage is entirely 'off-grid'; they have no electric or gas lines that connect them to larger-scale energy systems. Instead, Earthaven's residents produce their own electricity using individual photovoltaic systems and a cooperatively managed 'micro-hydro' system that generates electricity by diverting a portion of one of their mountain streams through a small turbine and back into the stream (Lockyer, 2007). Findhorn uses four community-owned wind turbines with a total generating capacity of 750 kW to produce 100 per cent of the community's electricity needs. These turbines are tied to the regional electric grid in such a way that they can export surplus energy to the grid when they produce an excess and import energy from the grid when the wind does not blow. In addition a Findhorn-based company supplies solar hot water panels both within and beyond community boundaries (Findhorn Foundation, 2009). In and of themselves, these technologies are not particularly novel. What is unique is that they are organized and utilized at a local, community scale by groups of people who conceive of themselves as experimentation and demonstration centres for ecologically sustainable models of energy production.

Even as ecovillagers experiment with alternative sources of renewable, reduced carbon energy, they also seek ways to decrease their levels of energy consumption by changing the way that their living spaces are designed. The International Energy Agency estimates that approximately one eighth of the total energy consumed by the average US resident is consumed in the process of heating, cooling, lighting and running appliances in the household (International Energy Agency, 2008). Per capita energy consumption has gone up as square footage has increased, the number of people per household has declined and home construction techniques have focused less on energy conservation measures in recent decades. The ecovillage movement counters these trends by building passively designed houses that take advantage of natural energy from the sun and air flows, are highly insulated and energy efficient, and increase the number of people per square foot.

For example, it is part of Earthaven Ecovillage's design code that individual houses are built to take advantage of passive solar heating opportunities, replacing heat produced by burning fossil fuels or wood with that of the sun's rays passing through appropriately oriented windows. Earthaven's residents are experimenting with a variety of technologies for holding that energy inside their homes. One household is based on the earthship model; it employs used tyres as wall structure and thermal mass to catch and slowly release the sun's energy over the course of the night. Other buildings, including Earthaven's main community building, use highly insulative materials such as straw bales to hold heat energy in their interiors. In addition, Earthaven's residents often bring multiple households together under one roof by creating unique combinations of public and private space (Lockyer, 2007). For example, one neighbourhood at Earthaven includes one building with a variety of separate family living areas, and offices, in combination with large kitchen and recreational areas that are shared by all occupants. This arrangement saves a portion of the fossil fuels that would otherwise be used in constructing and maintaining completely self-contained individual family dwelling units.

Reduced Consumption and the Construction of New Cultural Identities

Creating a life based on local food production, alternative currencies, bioregional economic organization and eco-technologies requires significant changes in daily practice and the cultural values and identities that guide behaviour. By coming together to support each other, the members of ecovillages like Earthaven and Findhorn are developing cultural alternatives to modern forms of consumerism with large carbon footprints. Consumerism has become a taken-for-granted way of life that is deeply engrained in contemporary, western cultural values and worldviews. Acts of consumption are not simply material acts or rational choices about what types of goods we need. Rather, they are tied up in broader social norms; our acts of consumption say something about our social identities relative to other people (Jackson, 2007). As a series of symbolic acts, consumerism is negotiable. In material terms, consumerism is highly fossil fuel dependent and thus contributes to our large carbon footprints. Finding a way to reduce the energy and material throughputs associated with consumer lifestyles is a necessary component of any programme to effectively mitigate climate change (Jackson, 2008).

Ecovillages contribute to the reduction of high carbon consumerism by supporting each other in developing alternative sets of cultural values that place an emphasis on sharing, cooperation and social relationships rather than conspicuous consumption. In ecovillages, people know and trust their neighbours and feel comfortable sharing goods such as vehicles, tools and appliances. Ecovillages create a context in which people support each other in their choices to emphasize social interaction and voluntary simplicity over conspicuous consumption and individualism. In turn, new cultural identities based on sharing and cooperation manifest positive climate change action because they reduce demand for consumer goods, in turn reducing the amount of fossil fuels consumed and carbon emitted in the production and use of those goods.

Several illustrative examples can be drawn from my field research at Earthaven Ecovillage. During my six months at Earthaven, I found that the length of time passed since one's last trip to the nearest town was a frequent topic of conversation. Statements such as 'I have not been to town for over three weeks' were often proclaimed with pride; conversely one's second trip to town during the current week was often accompanied by feelings of guilt for such reckless use of fossil fuels. At the same time, less frequent trips to town were facilitated by the sharing of items within the village. During my time there, one of the ecovillage members organized his extensive DVD collection as a lending library from which anyone could freely borrow. Such an arrangement was built on mutual trust and on the fact that one could be sure of seeing their fellow ecovillage members frequently enough that this trust would not be abused. At the same time, individual trips to town and the private consumption of material goods such as DVDs were much less often a focal point for people's free time than were potlucks (a variety of communal activities) or other social events during which social relationships were strengthened and shared values reinforced.

Similarly at Findhorn, social interaction and simple forms of recreation are emphasized as healthy and sustainable alternatives to consumer culture. A variety of places around the community are explicitly designed to serve as social hubs including the community store, the community centre which includes a large kitchen and eating space designed to feed hundreds at one time, and a community theatre where a variety of different forms of entertainment and community gatherings are regularly on offer. In addition, the community is designed to facilitate and promote access to and recreational use of the natural areas that the community has set aside as free from development. Preliminary evidence suggests that such social capital building strategies are not only good for the climate and the environment but for physical and mental health as well (Kasser, 2002; Mulder et al, 2006). Were all our communities explicitly designed to facilitate such social capital building while de-emphasizing conspicuous consumption, our collective carbon footprint might be significantly reduced.

TRANSITION TOWNS INITIATIVES

Many of the alternative models developed by ecovillagers are being translated into existing communities as neighbourhood groups come together to address the challenge of climate change. Transition towns initiatives bring together citizens of existing local communities and municipalities who are concerned about the threats posed by the decline of cheap oil supplies, changes associated with global climate change and the implications of both for our economic systems. The transition towns movement organizes people to address these problems by raising awareness and developing the skills needed to make local communities more economically resilient and self-sufficient, in the process reducing dependence on fossil fuels and the size of associated carbon footprints.

The transition towns movement started in Totnes, Devon, UK in 2005 and currently includes over 150 transition towns initiatives around the world (Transition Towns, 2009). The movement spread to the US in 2008, when a transition towns initiative was started in Boulder, Colorado. In 2009, a formal networking organization, Transition United States, was started, and within less than three months over 20 transition towns initiatives had been initiated in various communities from Los Angeles, California to Newburyport, Massachusetts (Transition Towns, 2009). The momentum generated in the process of mainstreaming and scaling up models gleaned in part from ecovillages is impressive, but perhaps not surprising given growing awareness of the impending consequences of unmitigated climate change.

Although the transition towns movement has explicitly designed a degree of flexibility into their definition of transition towns initiatives, they have laid out a series of steps for creating transition towns initiatives and a set of formal criteria that such initiatives must meet before achieving 'official' status. Setting up a transition towns initiative involves forming a group of committed community members who will develop practical tasks and skills for addressing peak oil, climate change and their economic implications. This group should aim to involve the broader community and local government representatives through inclusive, open space processes, regularly scheduled meetings, and the creation of committees focused on specific topics such as food, business, transportation and energy. Each initiative should develop a plan for the community's transition to a less fossil fuel dependent society. These plans should be concrete but flexible and designed to run their course in such a manner that the official initiative is no longer necessary and can quietly fade away once a plan is up and running in the local community (summarized from Transition Towns, 2009).

The formal criteria for achieving official transition towns initiative status, summarized below, are numerous. These criteria are designed so that the local community group has proceeded far enough in establishing a solid knowledge base, skill set and personnel group to ensure that the broader movement organization will not waste time and resources supporting a local initiative that does not have the ability to carry forward. Official transition towns initiatives are characterized by the following:

- an understanding of peak oil and climate change and their implications for the local community;
- a group of four to five people willing to fill leadership roles;
- at least two people from the core team willing to attend an initial two day training course;
- a strong connection to the local government;
- an initial understanding of the 12 steps to becoming a transition town;
- a commitment to ask for help when needed and work with the wider organization on funding applications;
- a commitment to regularly update your transition initiative web presence;
- a commitment to network with other transition towns;
- minimal conflicts of interest in the core team;
- a commitment to strive for inclusivity across your entire initiative;
- a recognition that although your entire county or district may need to go through transition, the first place for you to start is in your local community;
- at least one person on the core team should have attended a permaculture[1] design course (summarized from Transition Towns, 2009).

Of significance here is the movement's attempt to create a structure that is at once flexible enough to adapt to local conditions and rigorous enough to avoid the collapse and burnout that so often characterizes bottom-up movements. At the centre of this movement is the explicit emphasis placed on local communities as front line engagements with the challenges of climate change. Each transition town initiative is organized on a local, community level so that participants can interact with each other and create group strategies and projects that reduce carbon footprints and develop resilient local economies.

For example, Transition Town Totnes includes projects focused on the development of local food systems, helping local businesses transition to more sustainable energy use and operations, and creating a vision for a sustainable and fossil fuel independent Totnes of the future (Transition Town Totnes, 2009). With regard to local food systems, Transition Town Totnes has undertaken a number of different initiatives with the aim of working 'closely with the farming and business community to develop a relocalized food infrastructure and encouraging individual and community action to grow our own food' (Transition Town Totnes, 2009). To this end they have started a Garden Share project that links local garden owners who want to see their gardens become more productive with local gardeners who do not have regular access to garden space. In July 2008 over 40 people attended a local edible garden crawl and since then over 30 gardeners have become involved in the Garden Share project. Transition Town Totnes also produced a Guide to Local Food, Shopping and Eating that focuses on food producers within five miles of Totnes and the businesses that support them. They have developed a food hub that coordinates the desires of local food consumers and the abilities of local, small-scale, family-oriented producers. Finally, Transition Town Totnes has begun planting fruit and nut trees around their community with the aim of providing a long-term, nutritious food source for the community. This incomplete list represents only some of the ways that Transition Town Totnes aims to reduce the carbon footprint of their food systems while simultaneously generating a stable, sustainable and resilient local food system.

With regard to local businesses, Transition Town Totnes is undertaking a number of initiatives. In addition to producing a guide to businesses that are involved in local food systems, they also have an online list of local businesses that accept Totnes's local currency, the Totnes Pound. Reflecting previous explanations of the significance of local currencies, Transition Town Totnes (2009) explains that the Totnes Pound is designed to get money circulating in the local economy, build social relationships, get people thinking about the broader impacts of how they spend their money, and not least 'to encourage more local trade and thus reduce food and trade miles' and associated carbon emissions.

Transition Town Totnes is helping local businesses become more sustainable by educating them about energy efficient lighting and switching to green energy tariffs. The transition team offered free lighting audits to local businesses that demonstrated how much money and energy businesses could save by switching to energy efficient bulbs and provided access to a company that could implement the changes for them. Similar audits were offered with regard to business energy sourcing. The audits showed the potential savings to businesses of switching their tariffs to renewable energy sources and explained how the growing demand would lead to more competitive renewable energy prices and reduced carbon footprints.

All of these initiatives contribute to one of the main goals of every transition towns initiative – the production of an 'Energy Descent Action Plan'. Energy Descent Action Plans show how the local community can remain stable and even thrive without being dependent on high levels of fossil fuel consumption. Totnes's Energy Descent Action Pathways Project has a goal of reducing the current nine barrels of oil per person per year average down to one barrel of oil. Achieving this ambitious goal requires coordination among a wide variety of subgroups within Transition Town Totnes. As with so much in the transition towns movement, these initiatives are in their infancy and their impacts and outcomes remain largely unanalysed. However, the point here is that these initiatives are becoming increasingly numerous and are being started without prompting from governments on any level.

Transition towns initiatives' emphasis on working with local authorities is also a significant part of the overall vision of the movement. All groups that wish to form a transition towns initiative are encouraged to 'build a bridge to local government' not only to ensure that local authorities and policies don't pose obstacles to the strategies and projects being developed, but so that the transition towns initiative may start to exert an influence on local level policy-making through the construction of positive relationships with local authorities (Transition Towns, 2009). By working with local authorities, the transition towns movement aims to move their efforts to address the challenges of climate change beyond the arena of already convinced citizens and into the mainstream through local policy engagements. As the urgency of addressing climate change becomes apparent, the movement believes that local authorities will be grateful to have such forward thinking groups in their midst to serve as catalysts for broader change.

CONCLUSION: FACILITATING CLIMATE CHANGE ACTION IN LOCAL COMMUNITIES: OBSTACLES AND OPPORTUNITIES

The case studies and evidence reviewed in this chapter point strongly toward the potential of the voluntary and deliberate efforts of ecovillages and transition towns initiatives to serve as sources of data and positive models for broader climate change action. As the decades-long work of the Intergovernmental Panel on Climate Change makes clear, climate change is driven largely by the consumption of large volumes of fossil fuels. Fossil fuel consumption is a pervasive part of everyday life in most industrialized countries; our global trade systems, our food production systems, our household heating and cooling systems, residential and business electricity provisioning systems are all directly dependent on the burning of fossil fuels. Even our cultural values are implicated as the symbolic values of consumptive acts can have significant climate impacts.

Ecovillages and transition towns are citizen initiated endeavours that are trying to change this situation by redesigning both material and symbolic dimensions of the way we live while maintaining a certain level of comfort and satisfaction that make life meaningful and stable. This preliminary survey suggests that these initiatives have a lot of potential for success in the area of carbon reduction. This is consistent with a recent study conducted by the University of Kassel in which the per capita carbon footprints of two German ecovillages, Okodorf Sieben Linden and Kommune Niederkaufungen, were calculated and found to be, respectively, 72 per cent and 58 per cent lower than the German national average (cited in Dawson, 2007). Additional research is needed if the efforts of ecovillages and transition towns are to bear greater fruit. I would like to conclude this chapter by suggesting that scholars and government officials should actively develop partnerships with ecovillages and transition towns initiatives to conduct research and learn from their experiences as we confront the immense challenges of effectively mitigating and adapting to climate change.

Ecovillagers are increasingly conceptualizing their communities as living laboratories and demonstration centres for sustainable solutions to contemporary social, economic and ecological problems. Their desire to work with scholars, citizens and government officials is apparent in a recent GEN document, which says, in part, 'We see the principal mechanism for sharing our skills and models and for disseminating ecovillage values is via . . . developing partnerships with organizations and networks that are seeking to mainstream sustainability' (Global Ecovillage Network, 2008). The transition towns movement includes similar statements encouraging bridge building with local authorities and facilitating the participation of researchers in their endeavours. Despite such proclamations, scholars and politicians have been slow to develop partnerships with ecovillages and transition towns initiatives, especially in the United States. The development of such partnerships, if pursued with careful thought and design, represents an opportunity to use these existing 'living laboratories' as testing grounds and models for potential broader solutions while simultaneously enabling these groups to make greater progress toward their own goals.

The potential for conducting multi-, inter- and transdisciplinary research in ecovillages and transition towns initiatives is far reaching. Studies of the potential for economic relocalization, green building design and eco-technologies to reduce carbon footprints are only a beginning. Studies might also be conducted on the public health dimensions of intentional living and local agriculture. Economists might study the ability of local-ized economic systems to be resilient in the context of global economic shocks such as those recently experienced. Political scientists could study the dynamics of small group decision making and climate change action. Sociologists and anthropologists could study the genesis of new cultural identities and their transmission across cultural and generational bounda-ries. Engineers might use the experimental technologies of ecovillages as starting points for developing sustainable and replicable technologies that could also be scaled up. The list goes on and the natural laboratories where these experiments are happening – ecovillages, transition towns and other intentional community building endeavours – are already in existence.

The existence of and widespread public participation in ecovillages and transition towns initiatives indicate that many people want to take action to help mitigate climate change and that lacking appropriate opportunities provided by government authorities, they will create their own arenas for doing so. In the process, they often encounter existing policies and cultural attitudes that hinder their ability to continue taking the positive steps they envision. Building codes and zoning ordinances often force concerned citizens to abandon ambitious attempts to come up with new kinds of solutions that do not fit within the models for which policies were previ-ously designed. Creative attempts to build green, energy efficient housing, employ alternative energy technologies, and relocalize agricultural and economic systems often encounter existing policies that create significant obstacles – policies that were not designed to address the problems of climate change. Ecovillages and transition towns are important living laboratories and policies should be formulated to recognize their value as such.

Local governments can take action and devise policies to facilitate the experimental actions being undertaken in ecovillages and transition towns initiatives once they understand their value as living laboratories. At the very least, variances can be allowed and legal structures and tax incentives devised that will encourage and enable ecovillages and transi-tion towns to continue some of the experiments in economic localization, eco-technologies and cultural creativity described above. A much larger and perhaps equally important challenge is for governments to stop subsidizing corporate economic practices that promote and depend on conspicuous consumerism and the extravagant use of fossil fuels.

Absent the latter, governments could use policies to facilitate creative, community-based responses and strategies that manifest positive climate change and carbon reduction action. In this scenario, governments would become facilitators of positive community action that is already happening, often in spite of rather than because of government policies. All of this may sound utopian. Indeed, intentional communities such as ecovillages have long been dismissed as such. However, if there ever was a need for utopian solutions, the urgency of finding effective strategies for mitigating global climate change fits the bill.

NOTE

1. Permaculture is a paradigm for designing sustainable human settlements, communities and agricultural systems that was started by Bill Mollison and David Holmgren in Australia in the 1970s (Holmgren, 2002). At its core are sets of ethical precepts and physical design principles that guide the creation of systems for meeting human needs. Permaculture also provides an important foundational philosophy and skill set for many ecovillages.

REFERENCES

Assadourian, E. (2008), 'Engaging communities for a sustainable world', in G. Gardner and T. Prugh (eds), *State of the World 2008 – Innovations for a Sustainable Economy*, Washington, DC: Worldwatch Institute.

Bates, A. (2003), 'Ecovillages', in K. Christensen and D. Levinson (eds), *Encyclopedia of Community: From the Village to the Virtual World*, London: Sage Publications, pp. 423–5, Vol. 2.

Carr, M. (2004), *Bioregionalism and Civil Society: Democratic Challenges to Corporate Globalism*, Vancouver, BC: UBC Press.

Crate, S.A. and M. Nutall (2009), *Anthropology and Climate Change: From Encounters to Actions*, Walnut Creek, CA: Left Coast Press, Inc.

Dawson, J. (2006), *Ecovillages: New Frontiers of Sustainability*, Totnes: Green Books, Ltd.

Dawson, J. (2007), 'The path to surviving peak oil: the power of community', *Permaculture Magazine*, **54**, 42–5.

Earthaven Ecovillage (2005), 'Community service guidelines for resident supporting members and guests', Earthaven New Roots handbook, unpublished document.

Earthaven Ecovillage (2009), 'Earthaven Ecovillage: building a sustainable intentional community', accessed 24 May at www.earthaven.org.

Fellowship for Intentional Community (2009), 'Intentional Communities Directory: Fellowship for Intentional Community', accessed 15 May at http://directory.ic.org/records/?action=search.

Findhorn Foundation (2009), 'Findhorn Foundation: spiritual community, ecovillage and education centre', accessed 24 May at www.findhorn.org/index. php?tz=300.

Global Ecovillage Network (2008), 'GEN Manifesto', personal communication with Jonathan Dawson, president of the Global Ecovillage Network, June.

Global Ecovillage Network (2009), 'Search ecovillages: Global Ecovillage Network', accessed 15 May at http://gen.ecovillage.org/.

Holmgren, D. (2002), *Permaculture: Principles and Pathways Beyond Sustainability*, Hepburn, VIC: Holmgren Design Services.

Intergovernmental Panel on Climate Change (2007), 'Climate change 2007: synthesis report', accessed 27 September 2009 at www.ipcc.ch/.

International Energy Agency (2008), 'Worldwide trends in energy use and efficiency', accessed 18 September 2009 at www.iea.org/Textbase/publications/free_new_Desc.asp?PUBS_ID=2026.

Jackson, T. (2007), 'Sustainable consumption', in G. Atkinson, S. Dietz and E. Newmayer (eds), *Handbook of Sustainable Development*, Cheltenham, UK and Northampton, MA, USA: Edward Elgar, pp. 254–68.

Jackson, T. (2008), 'The challenge of sustainable lifestyles', in G. Gardner and T. Prugh (eds), *State of the World 2008 – Innovations for a Sustainable Economy*, Washington, DC: Worldwatch Institute, pp. 45–60.

Kasser, T. (2002), *The High Price of Materialism*, Cambridge, MA: The MIT Press.

Kirby, A. (2003), 'Redefining social and environmental relations at ecovillage at Ithaca: a case study', *Journal of Environmental Psychology*, **23**, 323–32.

Lockyer, J. (2007), 'Sustainability and utopianism: an ethnography of cultural critique in contemporary intentional communities', PhD dissertation, Department of Anthropology, University of Georgia.

Metcalf, B. (2004), *The Findhorn Book of Community Living*, Findhorn: Findhorn Press.

Metcalf, B. and D. Christian (2003), 'Intentional communities', in K. Christensen and D. Levinson (eds), *Encyclopedia of Community: From the Village to the Virtual World*, vol 2, London: Sage Publications, pp. 670–76.

Miller, T. (1999), *The 60s Communes: Hippies and Beyond*, Syracuse, NY: Syracuse University Press.

Mulder, K., R. Costanza and J. Erickson (2006), 'The contribution of built, human, social and natural capital to quality of life in intentional and unintentional communities', *Ecological Economics*, **59**, 13–23.

North, P. (2006), *Alternative Currency Movements as a Challenge to Globalisation?: A Case-Study of Manchester's Local Currency Networks*, Burlington, VT: Ashgate Publishing Company.

Raddon, M. (2003), *Community and Money: Men and Women Making Change*, New York: Black Rose Books.

Sale, K. (2000), *Dwellers in the Land: The Bioregional Vision*, Athens, GA: University of Georgia Press.

Transition Town Totnes (2009), 'Transition Town Totnes', accessed 24 May at http://totnes.transitionnetwork.org/.

Transition Towns (2009), 'Transition Towns WIKI', accessed 24 May at www.transitiontownsorg.

13. Energy Conscious Households in Action (ECHO Action)

Elliot Bushay

INTRODUCTION

ECHO Action was a two and a half year project part-funded through the European Union under the Intelligent Energy Europe Programme. It aimed to bring about direct involvement on a European scale of 2000 households in an active process of turning lifestyles and energy consumption patterns towards a model of sustainability. The programme was designed to help improve market conditions in support of renewable and energy efficient technologies across Europe, consistent with EU energy and climate change policy (EU, 2007a, b) through cutting greenhouse gas emissions and, in particular, reducing energy use from domestic household activity.

ECHO Action was run simultaneously in nine cities across seven European countries (Italy, UK, Sweden, Germany, Bulgaria, Lithuania and Portugal) between November 2006 and April 2009. The project was facilitated by local energy agencies in each city contributing to a network of partner organizations that developed local activities within the ECHO Action framework according to national and local energy policies and frameworks. The contribution from a diverse range of households, organizations and local authorities served to facilitate and initiate unique interactions on a broad scale, which led to important experience sharing and learning.

The ECHO Action project represents a bottom-up approach to addressing climate change issues at the community level, through focusing on households taking part in working groups and their attempts to realize a more responsible and sustainable use of energy and resources whilst travelling along a common path to change.

An important feature for a project of this type is the involvement of renewable energy and energy efficient technology suppliers alongside financial product suppliers in support of household decision-making. The project framework supports and encourages ongoing networking and information sharing between partner organizations and participants. This enabled partners, supporting organizations and households to experiment, introduce and share new and innovative ideas and techniques throughout the project, contributing to the organic growth and network of ideas, information and activity.

This chapter provides an overview of the ECHO Action project with an emphasis on the developments in London and highlighting common observations in groups from different partner countries.

COMMUNITY-BASED ENVIRONMENTAL ACTION PROGRAMMES

Community-based environmental action programmes appear to be a fast growing trend, with governments, local authorities, agencies and communities adopting and adapting variations of existing models for application to local circumstances.

Despite the relative success of programmes such as Global Action Plan's EcoTeams, Wiltshire climate friendly towns and villages, Ashton Hayes and the Transition Towns movement, a number of common challenges and issues exist; identified in past and ongoing projects and also recognized by partners in the ECHO Action project. One of the most important is the broadening of the socio-economic and cultural diversity mix of participants. Similarly, there are commonly shared concerns over the tendency that these initiatives have for attracting participants with pre-existing 'pro-environmental' mindsets. However, as pointed out in two large scale field studies conducted among representative samples of Dutch households, 'results showed respondents who indicate they behave pro-environmentally do not necessarily use less energy' (Gatersleben et al, 2002, p. 335). This statement was proven in the ECHO Action project on a number of occasions and is an area which merits further investigation. The issue is touched upon briefly later in this chapter when attention is given to the profiles of participating households.

For the purpose of this chapter, we will adopt the following definition as our understanding of a community-based action programme: 'a self-directed learning approach to bring together people who endeavour to share thoughts, ideas, feelings and concerns on climate change and to then take decisive action within their local communities' (Pierce, 2009, p. 1).

This definition is taken from a report on the progress of Climate Action Groups (CAGs) initiated by the Greater Reading Environmental Network and the Climate Outreach Information Network (COIN) located in Oxford. The programme followed a similar approach to ECHO Action in that members from all sections of a local community were invited to engage in an initial match making meeting where interests and ideas were communicated and developed over six subsequent meetings with ongoing support from COIN. The issues of participant 'type' and (a lack of) multicultural diversity stated above were apparent in the experience of these CAGs and highlighted in responses to participant questionnaires, which 'made reference to the majority of white middle class individuals in attendance . . . attendees were representative of the kinds of people who join low carbon communities' (Pierce, 2009, p. 3).

The observation raises an important question as to whether the current model and its existing variants is appropriate for engaging a broader spectrum of households across the UK in climate action at a local community level. However, even in this early stage of development and initiation of these types of networks, there are signs suggesting merit in their usefulness and appropriateness for mobilizing communities in climate change action and in reducing carbon emissions. If this is the case, then, a means of engaging all sections of society and effectively implementing such programmes on a larger scale will need to be identified.

Involvement among affluent and pro-environmental members of society is also recognized in one of the earliest programmes of community engagement on climate change issues, where a group of residents in Portland, OR took it upon themselves to apply a David Gershon blueprint to establish a CO_2 reducing campaign (Rabkin and Gershon, 2006).

The programme (a pilot) was greeted with considerable success, with a 43 per cent recruitment rate and the achievement of household-based CO_2 emissions reduction averaging 22 per cent. However, a key question left unanswered by this case surrounds the ability of a programme like this to attract similarly enthusiastic responses in areas where residents are less receptive to climate issues (Rabkin and Gershon, 2006). The report goes on to suggest that more tangible benefits need to be communicated to prospective participants in a much broader programme that include issues beyond the subject of climate change if larger audiences from a greater diversity of backgrounds are to be properly engaged.

This group of Portland households, and those taking part in ECHO Action, can be seen to represent the very few innovators as described in Rogers' (1995) five categories of adopters, namely 'innovators', 'early adopters', 'early majority adopters', 'late majority adopters' and 'laggards'. Rogers asserts that when an innovation is introduced, the bulk of people will either be early majority adopters or late majority adopters; fewer will be early adopters or laggards; and fewer still will be innovators (the first people to use the innovation) (Rogers, 1995). He then goes on to highlight the opportunity which exists for identifying the adopter characteristics of people which can be usefully drawn on to plan and implement strategies that are customized to their needs. In the light of this the current drive towards grass roots/community action and transition towns seems to be well placed for strategic planning in order to nurture and incorporate the early and late majority adopters.

THE ECHO ACTION APPROACH

The ECHO Action programme contains a number of features commonly associated with projects that engage with households at a community level. These include facilitating technical aspects and social dimensions in an attempt to nurture new and existing community relationships. The approach encompasses a number of dimensions set out by Defra's Market Transformation Programme (Jackson, 2005). As Jackson notes, 'The complex terrain of human behavior, as viewed in a social, psychological and cultural context, is not a place devoid of possibilities for state influence. Rather it is one in which there are numerous possibilities at multiple levels for motivating pro-environmental behaviors and encouraging sustainable lifestyles' (Jackson, 2005, p. 34).

Approaches were adopted under three of the four main headings illustrated in the 'New Model for Behaviour Change Policy' (HM Government, 2005; Jackson, 2005) and include: 'Enabling' – providing education and information with a view to taking steps towards removing known barriers to sustainable lifestyles; 'Engaging' – providing a platform for community action and utilizing personal contacts/enthusiasts and networks; and 'Encouraging' – through signposting available resources and grants and developing environments where recognition/social pressures and league tables are acceptable.

Another innovative aspect of the project was the involvement of suppliers of sustainable energy technologies, services and financial products. The involvement of these participants went beyond the traditional contribution from product delivery agents, in that suppliers and professionals in the energy efficiency field were invited to take on an additional educational and informative role through the provision of presentations followed by discussion with household groups. In this sense the programme facilitated a process of reciprocal learning.

In addition, a trainer was allocated to each group of households. The trainers accompanied participants over the duration of the project towards planning and implementing lifestyle and behaviour change. The trainers were the principal points of contact between participants, facilitating workshops and essentially acting as the bond between households, suppliers and other stakeholders, encouraging involvement and productive interactions. The trainers also provided additional information useful to participating households in achieving their desired goals and in support of local/specific interests and needs of the group in relation to borough-wide, national or European objectives.

WORK PROGRAMME

Recruitment

This section provides a summary of the recruitment process as applied to the recruitment of households in Northwest London. The same strategies were adopted across all European countries, adapted according to local circumstances. In cases where significant diversion or adaptation was implemented this is highlighted together with significant findings that emerged.

Households were selected to reflect the diversity and common features of communities in local areas; with groups and individuals targeted accordingly in order to ensure inclusion of a variety of social and ethnic backgrounds. Other criteria used to identify participants included:

- people living in a wide range of housing types;
- areas with a variety of good and less advantageous transport links;
- home owners, private renting or social housing.

Four main strategies were used to promote ECHO Action and invite households to take part in the project. The figures below refer to households recruited in the publicity campaign in Northwest London:

1. *Leaflets and posters*: Common marketing to members of the public through distribution of leaflets and posters across targeted areas. The difficulty with this approach, especially in densely populated inner city boroughs, is the overwhelmingly profuse amount of information that many citizens are already exposed to on a daily basis.
2. *Online and print media*: Publishing information in local newspapers and community websites. This approach proved more successful than the leaflets and posters, with approximately one third of participating households recruited through this route.
3. *Green fairs*: Approximately half the participating households were recruited at community or green fairs. The main advantage was that citizens attending these events are on the whole more open to environmental

and green issues and therefore more inclined to choose to commit to projects like ECHO Action. The theme, activity and atmosphere at such events geared towards green issues increase visitor interest.

4. *Direct mailing*: A fourth approach involved direct mail used in two ways. First by targeting individual households in specific streets or blocks of flats, which also allowed us to ensure a good representation of households taking part. Secondly, leaders of community groups, tenants' associations and similar organizations with a community focus were approached. This is seen as an efficient route to recruiting and working with households who share a range of common experiences or goals, increases the chances of group involvement, and proves logistically sound as most are likely to live or congregate in a local area. However, there is a strong reliance on the interest and motivation of group leaders, although, in cases where groups were recruited through this approach, the interaction worked well.

Despite this varied approach to household engagement there was a noticeable lack of interest from some targeted households. The main reason identified for this – apart from a common disinterest in engaging in pro-environmental behaviour – was unwillingness or an inability to commit to a 14-month programme. For the average household this is a substantial commitment. Although the time and energy involved are not significantly great (approximately two to three hours per month), when considered alongside the typically time constrained lifestyles of households living in inner cities, there is a level of personal application and concerted effort required for active participation. This goes beyond acting as a passive recipient of information apparent in many mainstream forms of awareness raising and behaviour change campaigns.

Working Groups

Working groups are formed from 'families' in a broad sense and so include cohabitants or communal households for example, where the objective is to create or reinforce a social network. In this sense, it is advantageous to involve entire blocks of flats or clusters of housing, in which collective decisions and actions are more feasible and effective, such as car pooling or ride sharing or the implementation of renewable energy systems on public spaces or buildings. The ideal is for group members to live no more than a quarter of a mile from each other, with participants reflecting a diverse mixture of households in terms of knowledge and opinions but also similarities with regard to their interests and needs in order to support collaboration and the exchange of ideas.

Referring to Global Action Plan's EcoTeams programme, Nye and Burgess (2008) point out that 'deeper analysis of the group-based mechanisms [in EcoTeams] shows that the group offers a complementary mixture of social support and pressure for making changes, and deliberative space for the diffusion and exchange of information and ideas for change' (p. 70).

The ECHO Action working groups followed a similar format to that used in EcoTeams in that each group was given as much freedom as possible in the planning, timing and content of workshops to develop a sense of ownership and to ensure interests were best catered for within the framework of the project. The formation of working groups presents the first opportunity in which community groups were galvanized through collective engagement in sustainability issues. From this point forward, two main ideas were presented and reinforced to group members. From the first workshops the initial idea was linked to a critical review of the individual participant's energy consumption and habits (outlined below). The second tied in the individual's personal energy use and behaviour with that of the group by highlighting the link between the contributions of emissions from energy use of individuals to that of the group as a whole. In this way an attempt was made to stimulate a collective awareness of personal and group consumption and subsequent emissions. Individuals contained within the working groups were encouraged to share findings from their reports, plans and actions throughout the duration of the project, offering support and sharing experience in a collective drive towards an agreed set of actions and sustainability targets.

On average there were approximately 12 participants per group. Group sizes would range from six to 20 members attending at any one time. Typically, the turnout per workshop would be between 25 per cent and 90 per cent of the original group; most participants would attend approximately 70 per cent of all workshops.

Workshops

Workshops were held over a 14-month period and are summarized by three successive levels of engagement:

- Level 1: critical review of consumption, behaviour and re-orientation of behaviour patterns;
- Level 2: implementation of easily achievable, low cost improvements;
- Level 3: substantial improvements to the home or modes of personal mobility or the realization of common interest actions (that is, the formation of purchasing groups).

In all cases there was a need to adapt workshops and work programmes according to the interest and needs of participating households. Workshops varied from group to group and country to country: from a more relaxed approach such as in Sintra (Portugal) where energy awareness was focused on at an early stage and workshops were informal and social in nature, to formal lectures as seen in Kaunas (Lithuania). To harbour a sense of collective purpose at an early stage of group formation, the first two workshops involved imparting information pertaining to climate change, local and national policies, drawing on current science and highlighting personal energy use amongst other factors as a route to making positive local contributions to help reduce individual environmental impact.

Our monitoring strategy involved two approaches. The first was aimed at households in a position to engage with the project on a deeper level and involved an energy audit and production of a report for the household after the introductory workshop. This was followed by monitoring home energy use (through recording meter readings) over an extended period of time. Energy consumption for each participating household for the year prior to taking part in ECHO Action was calculated and used as a benchmark for future planning and actions. The monitoring served to help households understand how much energy they were consuming periodically; enable them to identify and understand fluctuations in energy use, providing opportunities for improved energy management; and encourage them to set and aim for targets.

Not all households were prepared or in a position to engage in this level of monitoring over a prolonged period. Subsequently we asked households to gather energy consumption data at longer intervals, these being quarterly, at a mid-point and at the end of the project (depending on interests, needs and abilities). In this way it was possible to ensure that the majority of households were involved in the personal monitoring process. Calculations were made providing an indication of how well households were doing in terms of meeting personal and group targets.

A second approach to monitoring and predicting energy consumption involved the use of a monitoring tool, 'CO$_2$ Check', developed by ECHO Action partners for use during the project. The tool was useful for households unable to engage with the project as outlined above due to a lack of accurate data or joining a group late, for example. Households enter data regarding past energy consumption, behaviour, and number and types of lighting and appliances in the home, followed by an indication of planned actions over the coming year. The CO$_2$ Check tool is then used to calculate a notional energy saving according to the planned actions. This assisted households in understanding the potential effects of recommended changes.

In each case individual energy consumption was aggregated by group. Intermediate recordings of energy consumption were communicated to the group in terms of how well the group was progressing in relation to meeting the group target. Subsequent workshops involved households sharing findings from the audits and making commitments to attempt to reduce individual energy use over the coming year with a view to contributing to the collective energy reduction of the group according to current energy use and scope for behavioural and technical change. These initial workshops and exchanges set the foundation and scene for future interactions, with future workshops focusing on providing support and information in achieving household and group goals. A summary of the work programme is outlined in Table 13.1.

Companies

The aim of involving service companies (for example loft insulation, cavity wall insulation and renewable energy technology specialists) was to develop improved understanding of the interests, needs and capabilities of the households involved, thereby creating opportunities for new or improved products, services and relationships. The workshops presented an opportunity for companies and households to engage in a unique and informal experience and information sharing environment where neither party was subject to making (or expected to make) any formal commitment beyond a contribution to the workshop itself. Through presentations from suppliers and experts in specific fields, both customer and supplier had the opportunity to develop localized community networks in which local companies could offer the potential for households to utilize collective purchasing power.

The novel and ambitious category of 'ethical' or 'green' banks was included due to their significant potential to contribute to investment in renewable and energy efficient technologies either as direct suppliers of loans or as third party financiers of energy service companies. In each case this provided an opportunity to remove a well established barrier to investments in domestic energy efficiency in that 'households may not be aware of the long term savings (in financial terms or in comfort benefits) that result in investment in energy efficient products' (HM Treasury/Defra, 2002, p. 5). Through stimulating awareness on consumer home energy efficiency we sought to explore opportunities for mortgage lenders and other financial organizations to offer incentives for participating households to carry out recommendations highlighted in their ECHO Action home energy audit reports. This type of approach is encouraged by the Energy Saving Trust, which states that

Table 13.1   Example of a workshop programmme adapted for sample working groups

No.	Description	Aims	Objectives
1	Introduction	Establish mutual acquaintance, identify initial interests, needs and goals, take first steps to merging as a group	Outline issues and forthcoming activities, discuss and share thoughts, ideas, interests and goals
2	Data collection	Gain understanding of the auditing process and relationship to broader issues	Complete questionnaires, and discuss forms and types of information required to establish benchmarks
3	Findings and recommendations, where to start	Bonding – gain mutual understanding of where we all are and where we would like to arrive at, effects of action and inaction	Provide feedback from home energy reports. Agree on individual and group target for emissions over the coming months
4	The fabric of your home	Improved understanding of insulation technologies inspires action	Provide information, examples, of different insulation technologies and options
5	Heating – appliances and controls	Improve understanding of heating and controls to inspire action	Provide information, examples, of different heating and control options
6	Renewable energy technologies	Improve understanding of electrical appliances and renewable energy technology to inspire action	Provide information, examples, of different renewable and electrical related energy options
7	Transport	Improve understanding of travel behaviour and transport options, choices and support change	Issue or review travel diaries, provide information and resources in support of decision-making and change
8	Introduction to finance	Improve understanding of different finance options and opportunities	Provide information, examples, of different finance issues and options available

Table 13.1   (continued)

No.	Description	Aims	Objectives
9	Any other topic – group choice	Allow group to express ownership of workshops	Deliver workshop on the theme presented by groups
10	Update/review – making changes	Support households in need of support in making changes	Collect information and data and respond to specific individual requests
11	General meeting (social)	Support group bonding, allow group to organize and run an informal social event with environmental theme	Provide tools, information and resources for group to develop and plan and hold a social event
12	Summary, conclusion	Recognize and appreciate actions and achievements by all involved	Provide final feedback of project results and findings, look ahead to future activities

there needs to be a much more concerted effort by the major mortgage lenders to stimulate a serious focus on energy efficiency, and environmental issues in general. HM Treasury has a considerable degree of influence with the financial institutions, and promoting mortgages that encourage householders to reduce their carbon emissions would be a helpful move towards achieving some of the Government's own [targets for] energy and carbon savings. (Eppel, 2003, p. 24)

In each case (technological or financial), much longer term networking and partnership working is needed to develop financial networks and products, at a local level in particular, although some cities, countries and markets are better positioned to develop localized community networks that can incorporate financial institutes and technology suppliers. Moreover, involvement of companies requires a delicate balance to ensure the emphasis on local actions is maintained where small networks or clusters of groups can develop and retain an initiative which allows the use of collective negotiating and bargaining power; an important feature of the longer term development of community networks. The criteria for the 'economic' actors were to:

- operate in the local area;
- provide relevant technologies and services;
- show an interest or investment in operating in an ethical manner;

- be registered with a relevant governing body;
- be willing to provide products or services at a discount to ECHO Action participants;
- be willing to enter into a formal agreement (memorandum of agreement) or offer a statement/declaration of intent to support the project.

PROFILES OF PARTICIPATING HOUSEHOLDS

Three main types of data collection were adopted throughout the project in order to elicit qualitative and quantitative information which would help build profiles and understanding of individuals, groups, and their relationships and behaviour.

Four questionnaires were developed and utilized throughout the project. The first was used to collect very general and broad information from individuals expressing an interest and/or signing up to take part in the project at events or through contacting ECHO Action by telephone or email. The questionnaire requested information regarding the participant's age, location, age and type of property, interests in the project, and a brief overview of perspectives on climate change and any actions taken in the recent or distant past. The second questionnaire at the first workshop would then provide an opportunity for the individual to expand and provide a more detailed account of their personal energy use, understanding and perspectives on related issues. Questions regarding expectations from participation in the project with regard to what the participant expects to achieve and contribute also allowed us to plan, monitor, motivate and evaluate participant activity throughout the project. Mid-point and end-of-project questionnaires were then completed to support progress and help generate a rounded analysis.

Each workshop lasted approximately one and a half to two hours and was typically divided into three sections. The first involved presentation of information on a specific topic, solar water heating for example, which was followed by a period of feedback and discussion on the topic. Participants were encouraged to present their understandings, concerns, experiences and perspectives. In this way learning from the expert and the experiences of non-experts contributed to the group learning as a whole. These discussions provided the facilitator with opportunities to collect valuable information through observation of indirect exchanges and through prompting for deeper responses to emerging topics.

> The lifestyle concept does not only refer to the formal process of integration of social practices but also to the 'story' which the actor tells about it. With each lifestyle there is a corresponding life story, in the sense that by creating this

specific unity of practices the actor expresses who he or she is or wants to be. (Spaargaren and van Vliet, 2000, p. 55)

We received a 98 per cent take-up of offers for home energy audits, used in part to incorporate one of the 'principles of reciprocation' (Seethaler and Rose, 2006). Besides collecting technical information about properties and consumption, the time spent during the audit allowed for more valuable interpersonal exchanges (in the comfort of the home) where participants were more likely to be 'tuned in' to the range of everyday positives and nega-tives that influence energy consumption, behaviour and habitual routines.

> It is well documented that focus groups, whilst providing a good understanding of public opinion, are not always representative of 'real-life' actions, for example people who are thoroughly supportive of installing insulation or renewables in a focus group may be constrained back in their homes by a variety of factors that it is impossible to 'un-pick' in a group setting, for example, cost, family objections, time, knowledge etc. (Brook Lyndhurst, 2007, p. 4)

Thus informal discussion and observation during energy audits allowed for further development of insights into the life story of the participants, in particular with regard to intentionally oriented environmental behaviours (significant from the viewpoint of the actor) and impact-oriented envi-ronmental behaviours (actual environmental impact) according to annual energy consumption in this case (Stern, 2000).

Data collected from participants through questionnaires, discussion, audits and observations were categorized with reference to Defra's frame-work for pro-environmental behaviours (Defra, 2008), which considers society in relation to seven clusters according to attitudes, beliefs and behaviours. In this way we were able to generate profiles based on qualita-tive, quantitative and desk-based research.

Overall, five broad profiles were identified among households taking part in the project which bear similarities to the Defra model: 'Green con-sumers' (of which Defra's equivalent profile is called 'Positive Greens'), 'Consumers with a conscience' ('Concerned Consumers'), 'Energy savers' and 'Practical needs focused consumers' ('Waste Watchers'), and 'General interest' ('Sideline Supporters' and 'Cautious Participants'). The profiles are briefly summarized as follows:

Green Consumers

These participants demonstrate a very positive attitude and understand-ing towards environmental issues. There is a high level of commitment to aligning lifestyles to models of sustainability through consciously connecting understanding of the environmental impact of personal energy use to wider environmental issues and applying principles of sustainability. Households of this profile are likely to be active politically either through membership of political organizations or through supporting green lobbies. Commitment to workshops is dependent on the level of time available to take part actively and the contribution that can be made to the project and workshops as opposed to the amount of new resources or information that can be gleaned from participation. On the whole, input and involvement from these participants are high despite the limitations in the levels of new information available.

Consumers with a Conscience

These participants expressed similar levels of knowledge and understanding of environmental issues to green consumers, and expressed the same connections between lifestyle/behaviour and wider environmental issues. However, members of this group are likely to have identified greater restrictions – personal, financial, practical – than the typical green consumer and see them as distinct barriers to achieving sustainable lifestyles. Typically, participants of this type would express a need to 'do what we can' to sustain or protect the planet or environment. Again, there is a high level of motivation among these participants in terms of the willingness to take action and actions taken. The main motivation here is a concern for the environment with understanding and concern for personal contribution to environmental issues. Participants of this type demonstrated a high level of commitment to ECHO Action.

Energy Savers

Energy savers represent those participants adopting a pragmatic approach to environmental issues whose focus is primarily on energy efficiency. These participants placed environmental issues behind cost savings and energy efficiency and were less likely to identify themselves with green ideas. Recognition of achievements and actions is however important to this group, who also express keen interests in sharing and expressing knowledge and achievements to the group.

Practical Needs Focused Consumers

In the context of ECHO Action, this participant can be seen as transient. Typically, an existing need or interest has been identified (heating replacement, maintenance, controls, insulation or renewable energy) by the household, and projects such as ECHO Action offer an opportunity to meet this need at a reduced cost from a trusted source. These participants typically express little or no obvious interest or concern for environmental issues, and are more likely to reject arguments and be sceptical with regard to the validity of the science behind global warming for example or the motives behind government, industry and organizations for enunciating such issues and maintaining them in the public eye. Typically, this participant loses interest in projects once the need for information, products or services has been met. There is usually a high demand for high quality products, services or information, and providing it is practical, economic and achievable, the product is most likely to be invested in.

General Interest

This category represents participants who, prior to taking part in the ECHO Action, may not have expressed any specific environmental interests or ambitions. In addition, technical interests beyond the normal minimal requirements for understanding and engaging with energy efficient issues are a low priority. However, stimulation through awareness raising campaigns and general media, interest and ambitions may emerge where there is a desire to enhance existing knowledge and experience to a level beyond that of general common knowledge. This interest can become manifest and satisfied in many forms. Through taking part in workshops, even with no prior technical or environmental knowledge, participants of this category appear to seek to be a part of or involved in groups which can offer a sense of belonging and opportunities for learning and socializing.

Figure 13.1 illustrates summary profiles of individuals taking part in the two ECHO Action groups operating in London. In each European partner city a similar mix of participants was identified, with the majority of participants represented by 'green consumers' and 'consumers with a conscience'. Of the initial households, 60 per cent maintained an interest and stayed committed to the project. Of those participants who were able to stay involved, it was those who were members of existing groups such as church or other community groups whose numbers were the most stable and consistent throughout the project. Of the participant types who dropped out, it was those with practical needs, some greens and general interest where the greatest numbers were lost. On the whole, consumers with a conscience and energy savers made up the bulk of the participating households throughout the duration of the project. The primary reasons that participants gave for not maintaining involvement in the project included 'limited time to attend workshops', 'inappropriate workshop content' and 'impractical location of workshop'.

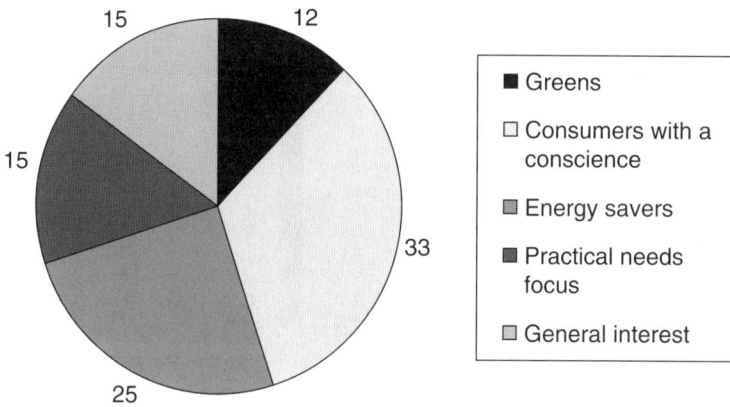

*Figure 13.1   Summary of participating households in the two London
ECHO Action groups by profile categories per cent*

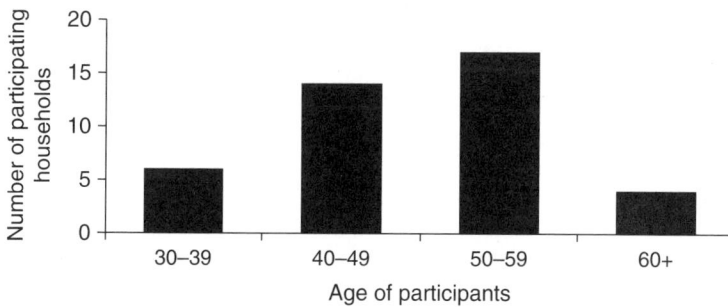

Figure 13.2   Summary of key group participants by age

The majority of participants from the two London groups were home-owners and, as can be seen from Figure 13.2, large proportions were between the ages of 40 and 59, with the majority from 50–59.

PRINCIPAL FINDINGS, MONITORING AND RESULTS

Of the participants taking part in the initial monitoring, approximately 60 per cent continued to monitor their own consumption and provide feedback. Data collection is a useful and valuable method of measuring, monitoring, highlighting and communicating progress and change to households, and linking in data with group and wider activities gives greater meaning and a

Low carbon communities

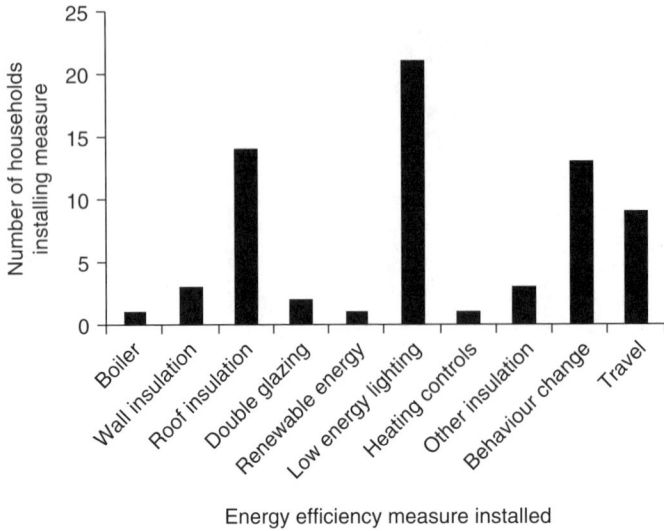

*Figure 13.3   Summary of energy efficiency measures installed three
quarters of the way through the first year by sample group*

more robust contextual significance. The interest and anticipation from the
majority of group members were particularly noticeable (for example, 'it's
like waiting for exam results!') and most were keen to be seen to have made
a positive contribution to the collective target of the group. This attitude
was particularly evident amongst households falling into the Consumers
with a conscience and Energy savers participant profiles.

Figure 13.3 presents a summary of measures installed by participants
in the two London groups having completed the third quarter of the programme. Important factors to note here are:

1. A large proportion of households involved at this stage fall into the
   category of those who are 'energy aware' and, by extension, already
   comparatively active in terms of home energy efficiency. As a result a
   number of energy efficient measures would have already been installed.
2. There are technical and economic barriers preventing some households from implementing certain desired measures.

In Figure 13.4 a summary of the measured changes in energy use from a
sample of nine households in the two London groups is presented, which
provided complete data (from energy bills) at the start of the project and
presented consumption data at the end of the third quarter. The average

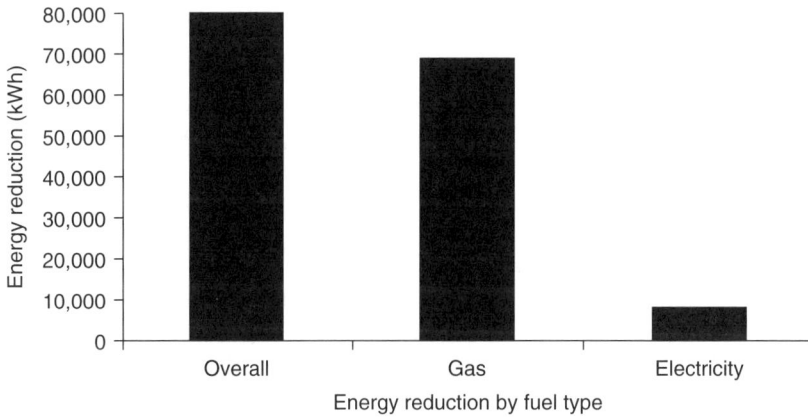

Figure 13.4   Summary of energy reduction achieved by sample groups

energy consumption for the sample groups is 23 351 kWh for gas and electricity combined. The figures have been adjusted to take into account the shorter length of time for the monitoring period displayed and any significant changes in household occupancy. Adjustments that account for differences in climate are made at the end of the annual period.

As described earlier, the figures above do not tell the full story of the groups involved and their activity and achievements, but are useful as an indicator. Factors such as the scope for actually achieving a reduction in energy use through behaviour as well as physical changes to homes, for example, play a significant part in the final results from the individuals and working groups. Of particular interest is the level of reduction achieved in energy use from the use of gas compared with electricity consumption. This is attributed to a number of factors, including the much greater scope to reduce gas consumption due to the comparatively greater quantities used, and the existing incentives available to the sample group through insulation schemes. However, at this stage, the average percentage reduction in energy use from the sample groups is approximately 11 per cent for both gas and electricity. It is expected that significant longer term savings will be achieved once insulation measures have been in place and monitored for a full year.

SUMMARY AND CONCLUSION

ECHO Action presents an ambitious approach to engaging households in a community-based initiative to tackle climate change issues, which contains value beyond that of initial carbon savings, and supports past and existing studies and work by environmental organizations in high-lighting the value of community action.

However, there are areas in this approach which suggest that more needs to be done to attract households and individuals from a broader range of societal sectors. Engaging 'environmentally unconcerned' households as well as younger members of the community remains a core challenge that needs to be addressed in order to establish stronger community awareness and action. The consensus from partners working on the project is that there is scope to extend the project or the key features on a much wider scale. This raises the question as to how diverse the households taking part can be and also how the programme might be tailored to suit different groups and individuals. Importantly there is little doubt about the interest and motivation generated and shared in groups from the majority of par-ticipants once they are involved. A key sticking point is the length of the programme. Careful timing (highlighting the relationship between recently introduced energy performance certificates and the content of topics covered in the project for example) and targeting of individuals, groups and social networks – in addition to tailoring of future work programmes – all hold the potential for substantial progress to be enabled.

As an example, partners in Italy reported a strong need for and interest in technical assistance and re-assurance from an impartial source such as the energy agency in particular when planning higher cost installations such as solid wall insulation or renewable energy technology. In Berlin and Bourgas (Bulgaria), a strong targeted approach was used which proved very effective. One and two family owner-occupier households and multiple occupancy buildings respectively were the primary criteria used to identify suitable areas and households with which to engage. In Bourgas this gave the energy agency the opportunity to work with some of the harder-to-reach households living in social housing through working closely with the local authority to engage with households living in blocks of flats undergoing renovation.

The importance of 'champions' cannot be underestimated in terms of galvanizing groups and sustaining interest in projects towards the stimula-tion of community action. In London, in particular, emphasis was placed on working with existing groups, such as religious, social or community groups, where valuable networks and relationships were already well established. Where champions are identified (group leader or member) and good relationships can be established with the trainer, the formula for success is significantly strengthened. This points to the scope for greater success through working with existing community networks, introducing and focusing attentions towards matters of sustainable lifestyles rather than attempting to create new 'sustainable communities' *per se*, the challenge being to identify areas of sufficient importance and interest where energies can be focused.

Contributions from specialists and companies are a very positive element but the level of value is country or city specific. There are stark contrasts in the availability and up-take of company discounts in Venice and Bologna where a high demand for technologies and services amongst households exists. In London, existing schemes supported by utility companies already provide households with very good discounts – particularly for basic insulation measures – meaning there was little room or incentive for local insulation companies to compete. In addition, smaller local companies are unable to compete with the subsidized products and services in this area.

To the extent that the fundamental levers for change have been identified in the majority of participating households, the concept could well be replicated, or elements incorporated, into new or ongoing projects at progressively larger scales, using common methodologies and tools. Important is the long term development of group activities beyond the initial engagement. The experience of ECHO Action serves to demonstrate that interests and ideas can – and do – develop at later stages and that the duration of planned actions may not necessarily fit into the time scale initially designated for the project.

ECHO Action represents a foundation for the longer term engagement of individuals and household groups. This could, in time, be developed on a much larger scale through continued support for the generation and maintenance of strong local networks where a common goal, once identified and understood, can lead to well established long term movements towards low carbon communities.

REFERENCES

Brook Lyndhurst (2007), *Public Understanding of Sustainable Energy Consumption in the Home: Final Report to the Department for Environment, Food and Rural Affairs*, London: Department for Environment, Food and Rural Affairs.

Defra (Department for Environment, Food and Rural Affairs) (2008), *A Framework for Pro Environmental Behaviours*, London: Defra.

Eppel, S. (2003), *Fiscal Incentives for Home Energy Efficiency*, submission to HM Treasury/Defra Consultation on Household Energy Efficiency, London: Energy Saving Trust.

European Commission (2007a), *An Energy Policy for Europe*, communication from the Commission to the European Council and the European Parliament, Brussels: Commission of the European Communities.

European Commission (2007b), *Limiting Global Climate Change to 2 degrees Celsius: The Way Ahead for 2020 and Beyond*, communication from the Commission to the Council, the European Parliament, the European Economic and Social Committee of the Regions, Brussels: Commission of the European Communities.

Gatersleben, B., L. Steg and C. Vlek (2002), 'Measurement and determinants of environmentally significant consumer behaviour', *Environment and Behaviour*, **34** (3), 335–62.

Gershon, D. (2009), *The Empowerment Institute*, accessed at www.empower-mentinstitute.net/files/SLC.html.

HM Government (2005), *Securing the Future: The UK Government Sustainable Development Strategy*, Norwich: HMSO, accessed at www.sustainable-development.gov.uk/publications/pdf/strategy/SecFut_complete.pdf.

HM Treasury/Defra (2002), *Economic Instruments to Improve Household Energy Efficiency: Consultation Document*, London: Her Majesty's Treasury and Department for Environment, Food and Rural Affairs, July, accessed at www.hm-treasury.gov.uk/d/household_energy.pdf.

Jackson, T. (2005), *Lifestyle Change and Market Transformation: Briefing Paper Prepared for Defra's Market Transformation Programme*, London: Department for Environment, Food and Rural Affairs.

Nye, M. and J. Burgess (2008), *Promoting Durable Change in Household Waste and Energy Use Behaviour: A Research Report Completed for the Department for Environment, Food and Rural Affairs*, Norwich: University of East Anglia.

Pierce, O. (2009), *Evaluation of Reading Climate Action Groups*, accessed 18 August at http://coinet.org.uk/what-we-do/climate-action-groups.

Rabkin, S. and D. Gershon (2006), 'Changing the world one household at a time: Portland's 30 day program to lose 5000 lbs', in S. Moser and L. Dilling (eds), *Creating a Climate for Change: Communicating Climate Change and Facilitating Social Change*, Cambridge, UK: Cambridge University Press, pp. 292–303.

Rogers, E.M. (1995), *Diffusion of Innovations* (4th edn), New York: Free Press.

Seethaler, R.K. and G. Rose (2006), 'Using the six principles of persuasion to promote travel behaviour change: findings of a TravelSmart pilot test', *Road and Transport Research*, **15** (2), 95–107.

Spaargaren, G. and B.J.M. van Vliet (2000), 'Lifestyles, consumption, and the environment: the ecological modernisation of domestic consumption', *Environmental Politics*, **9** (1), 50–77.

Stern, P.C. (2000), 'Toward a coherent theory of environmentally significant behavior', *Journal of Social Issues*, **56** (3), 407–24.

14.   The HadLOW CARBON Community: behavioural evolution in the face of climate change

Howard Lee and Julie Taylor

INTRODUCTION

Hadlow village is situated approximately 5 km from Tonbridge and 15 km from Maidstone in the west of the county of Kent, UK. It is a village of 1851 houses and is situated in a rural part of the county within a 'green belt' area.[1] Hadlow village has a long history of occupation, being known to have been settled by the Romans and continuously thereafter, chiefly as a centre of hop production for the brewing of beer. Today the village is known for its distinctive 50-metre-tall Hadlow Tower, a surviving portion of a demolished late Victorian Gothic structure.

Hadlow College is situated close to the village, covering a total of 286 hectares. The main campus is adjacent to the southwestern side of the village, with glasshouse facilities on the eastern edge and a dairy farm to the southeast. The college was developed in 1967 as a centre for land-based training, primarily in agriculture and horticulture but more recently as an important centre for other subjects such as equine studies, animal management, fisheries, landscape management, floristry, garden design and countryside management. Hadlow College currently teaches 1361 Further Education (aged 16–18) students and 548 Higher Education (degree) students, with short courses and evening classes attended by more than 2000 adult learners.

This chapter reviews the activities of a low carbon community initiative established in 2007 as a collaborative venture between Hadlow College and the village. Whilst many low carbon community initiatives have been formed in the UK in recent years, no others are known to encompass such a link. The mission of the aptly named HadLOW CARBON Community has been to bring village residents and College staff and students together and 'walk the talk' against climate change. This has involved some extraordinary projects, which have tapped into the considerable knowledge, skills and experience of people within both the village and College. Reflection on progress so far has highlighted the need for trust and cooperation for mutual benefit – the classic definition of social capital – which is discussed in this chapter.

The threats from climate change are so serious and pressing that local vision and community action are needed, to catalyse similar developments elsewhere. Thus, imaginative and successful low carbon community initiatives such as the HadLOW CARBON Community can rapidly inspire wider moves towards climate change mitigation and adaptation. How this developed at Hadlow and the lessons that can be learned for other projects will be discussed.

THE HadLOW CARBON COMMUNITY

The concept of a low carbon community initiative for Hadlow was first mooted at a Parish Council meeting on 25 April 2007 and received instant and widespread support. The first public meeting was organized on the evening of 28 June 2007. College students helped to deliver leaflets in the village and posters were widely displayed. The attendance was modest but there was much enthusiasm to form an active group. A better attended meeting was soon arranged and resulted in the formation of the HadLOW CARBON Community and a steering group to drive ahead with useful ideas. This chapter reviews: (i) two visionary and original projects – a 'Grower Group' and an art exhibition – initiated by the HadLOW CARBON Community; and (ii) what can be learned from this for the benefit of low carbon community projects elsewhere.

THE HadLOW 'GROWER GROUP'

This project began after a series of steering group meetings in late 2007 and early 2008 when the concept of local food production was discussed. HadLOW CARBON Community members wanted to undertake something novel that could demonstrate practical action against climate change and move towards sustainable development. In a low carbon future it was accepted that most if not all food would need to be produced locally, to minimize transportation impacts. Furthermore the changing climate was seen as a potential challenge to future food production, due not just to rising temperatures in the summer but increasingly severe extreme weather events throughout the year (Salinger et al., 2000). The steering group considered that low carbon communities could take greater responsibility for securing their own food supplies in a more uncertain future and that the HadLOW CARBON Community should explore how this might be achieved.

The term 'Grower Group' was coined, to denote a group of people who worked together to plan, plant, maintain, harvest and share food produced on one collective plot. Other terms have also been used for similar projects, such as 'community supported agriculture' by organic farmers and 'community allotments' by the Transition Towns movement (Howe, personal communication). In 2008 the Hadlow College Senior Management Team agreed that a plot of college land should be made available for Grower Group development. From early 2009, College Countryside Management students began installing rabbit-proof fencing around the plot, measuring 50 by 25 metres, as part of their curriculum training and Grower Group members started work soon afterwards. There has been an initial commitment from 12 local families in Hadlow for 2009, coordinated by HadLOW CARBON Community member Nicola Canham.

The latest version of the ongoing plan for the Grower Group plot can be seen in Figure 14.1. There is a clear balance between productivity and aesthetic appearance – note the circular, central flower bed. Car tyres and embedded wine and beer bottles are also used elsewhere to create smaller ornamental features planted with annual flowers. Vegetable production is currently focused on raised beds and Mypex-mulched beds (Figure 14.2). Both these production methods have clear advantages: raised beds are an ideal way of concentrating and maximizing organic matter and helping to retain moisture in droughts whilst also protecting plants from floods; Mypex-mulching also helps retain soil moisture and additionally smothers most weed growth, reducing the need for hand weeding or herbicides. Further details of production techniques are not presented here, since they are well documented elsewhere (for example, Pears, 2004).

The social organization of the Grower Group has been important for success, but has so far met challenges in terms of the number of active participants. While 18 people originally expressed interest in the project and 15 were initially involved, a core group of nine to ten volunteers has persistently worked on the site. Staff and student involvement has so far been confined to curriculum activities, such as the development of a rabbit-proof fence around the perimeter and construction of raised beds. Discussions with Canham indicate that the Grower Group needs more members to spread the workload and develop the site effectively. Increasing hardship in the current economic recession suggests that additional people may join as a means of securing cheaper vegetables but further efforts to recruit new members are also underway.

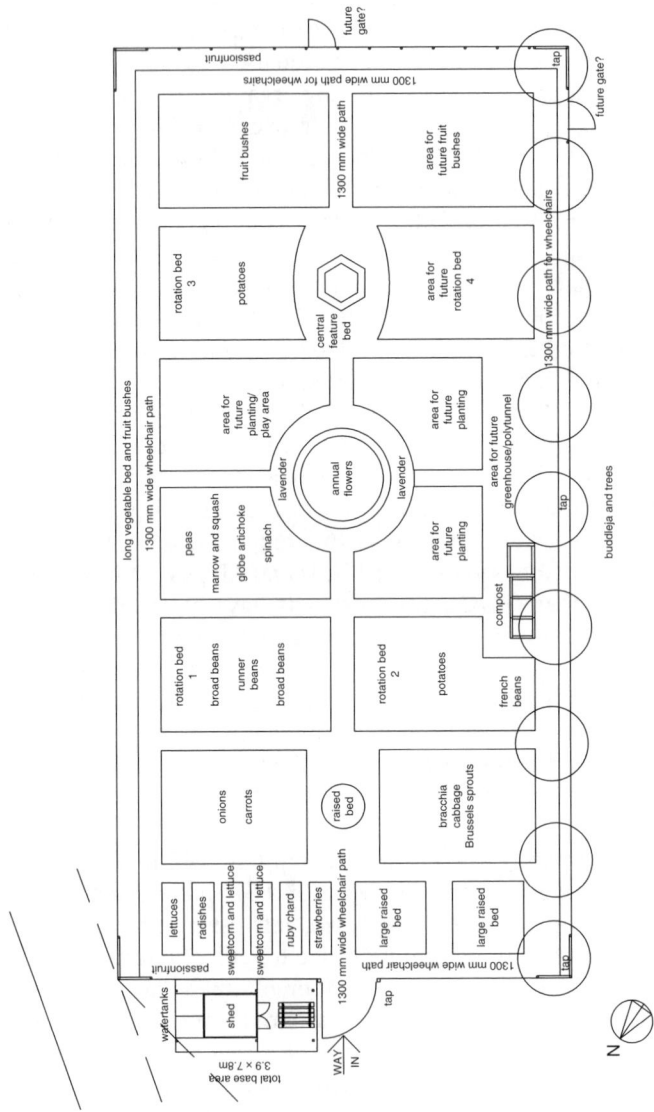

Note: Scale approximately 1:200 (at A4).

Source: HadLOW CARBON Community.

Figure 14.1   HadLOW CARBON Community Grower Group plot plan

(b)

(a)

Source: HadLOW CARBON Community.

Figure 14.2 Vegetable production in (a) raised beds and (b) Mypex-mulched beds

241

Positive Outcomes of the HadLOW CARBON Community Grower Group Project

Operation of the project is structured around a series of core principles designed to maximize positive outcomes. These include:

- Commitment – the need for both college and village stakeholders to commit to a Grower Group project for at least five and preferably ten years or longer. This gives security on both sides and especially means that Grower Group participants can invest time and effort into the site knowing that the college will be supportive, irrespective of staff changes. This commitment on both sides must be transparent and open, and supported by a joint management committee.
- Communication – should be optimized within-college between all strands of curriculum and estates/maintenance, but also between college and the Grower Group. While the college is ultimately responsible for all aspects of site management, it should ensure that Grower Group members are fully informed, supported and guided appropriately, using clear, jargon-free language. This support should include first aid cover for Grower Group members when on site and access to basic amenities. A 'College Champion' needs to be allocated and an opposite number in the Grower Group, to coordinate links via the joint management committee.
- Trust – the least tangible but most important criterion, which can take years to fully develop, and can be fragile and vulnerable to misunderstandings. Trust must be nurtured and patience and tolerance shown on both sides if the project is to succeed.

If a Grower Group organizes a plot of land in collaboration with another organization such as a farm or a council, then many of the above core principles will still be valid.

THE 'FINITE' ART EXHIBITION

This project was the initiative of local HadLOW CARBON Community artist member Julie Taylor. Taylor had been seeking a new way to express her concerns about climate change through her art. This approach has been more recently considered by Daniels and Enfield (2009), who review cultural responses to climate change, including an exhibition which resulted from journeys by artists to the Arctic (pp. 7–8). Earlier work by Gablik (1997) was particularly inspirational, regarding the effects of climate change on western society artists and what responses they might take. For example, one suggestion by Gablik is to disregard the western ego of the solitary artist and move towards more ancient ways of expression: 'art became very much ego-centred . . . whereas in India, art comes from the community' (Gablik, 1997, transcript of interview of Satish Kumar, p. 144).

In essence, therefore, the artist can surrender the ego of art ownership in order to engage in a deeper dialogue with the community, becoming the initiator of a project collectively produced and owned by community participants. Taylor discussed a range of ideas with six other artists and they became the core group that organized the Finite Exhibition. The poster design for this is shown in Figure 14.3. It was planned to develop the theme of climate change by means of tapestries, sculptures, paintings, etchings, moving-image sequences and weavings. Hadlow College Senior Management Team offered the HadLOW CARBON Community the Garden Design Centre on campus as a venue and Kent County Council also agreed to offer financial support. This eventually enabled a total of 37 artists to become involved, 14 relevant local organizations, including seven from Hadlow village, and two community artworks programmes.

Two very different approaches proved successful for the exhibition:

1. Art works controlled by the artists, such as the 'Sock Weave' designed by Elizabeth Cousins to demonstrate the power of reusing fabrics in conceptual weaves. For this exhibit, socks had been collected via the local school and church. At the exhibition visitors were able to participate in the weave, though it was ultimately a pre-determined, artist-controlled design.
2. Art works with audience participation, such as a blue coloured 'River Knit', constructed of knitted rectangles, of varying lengths, to represent flooding due to climate change. This project was the idea of Stephanie Ingham, a local architect and current chairperson (at the time of writing) of the HadLOW CARBON Community. Ingham had been inspired by seeing an artist producing a blue (squares) knitted 'river' at The Eden Project (2009) and wanted to develop this theme at the Finite Exhibition. The process of active participation enabled by 'River Knit' has clear resonances with Gablik's discourse on engagement in artwork where egos are subsumed, allowing the project to evolve into something less planned but equally valid and worthwhile (Gablik, 1997).

As Ingham facilitated the communal evolution of the project, this concept allowed new ideas to be introduced by varying sections of the community in an unplanned manner but which led ultimately to a constructive outcome. Villagers, Hadlow College students, people from the wider area

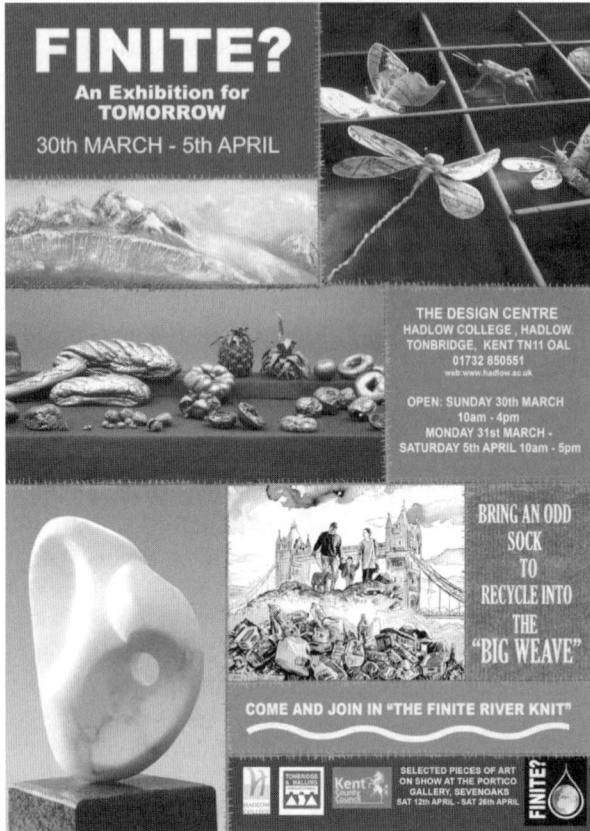

Source: Finite? Artists.

Figure 14.3   Finite Exhibition poster

and visitors to the show all contributed, and the 'river' quickly changed out of all recognition, with additional colours added to symbolize other climate change conditions such as drought. A picture is shown in Figure 14.4.

The Finite Exhibition was held in early April 2008 and was very much a leap of faith by both the college and village art community. Both sides were slightly nervous about the final appearance of the exhibition and its likely reception by local people and the media. The final outcome was hailed by all as a major achievement and a great success. Local TV and newspaper coverage was extensive and feedback was enormously positive: the use of art to spread the climate change message had clearly been highly

Source: HadLOW CARBON Community.

Figure 14.4   The River Knit in its early stages, with some contributors

effective. The HadLOW CARBON Community, within which the exhibition had been organized, showed real collaborative effort: Taylor worked virtually full time as the coordinator in the three months leading up to the opening and was supported by artists who donated and set up their work, and many volunteers who staffed and supervised the exhibits. The college was also involved, with estates staff supporting the development of the venue and other personnel assisting in the overall planning and organization, such as invitations and publicity. Nearly 200 people attended the celebratory opening night, including college staff, students, governors and local councillors and artists.

Overall, the Finite Exhibition involved a high level of collaboration between the college, HadLOW CARBON Community and Kent County Council. On reflection this was felt to have contributed to a more positive message; that is community action on climate change gave hope for the future. This more upbeat message led to additional benefits, including fresh interest and enquiries about the Grower Group.

DISCUSSION: WHAT CAN BE LEARNT, FOR THE BENEFIT OF LOW CARBON COMMUNITY PROJECTS ELSEWHERE?

The above two projects illustrate a profound shift in cognitive response by Hadlow stakeholders to climate change. Previously there had been

comments from college staff, students and village residents about feelings of helplessness engendered in the face of this issue. Subsequently,
the formation of the HadLOW CARBON Community has empowered
members to embrace positive action: there is now a collective feeling that
useful change is both possible and meaningful. This development will now
be discussed.

The Paradigm Shift at HadLOW

In 2007 there had been informal discussions and seminar workshops with
most staff, students and local villagers who attended the open HadLOW
CARBON Community meetings and the first few steering group sessions.
All college and local people were well aware of the concept of climate
change but many argued that they felt powerless in the face of such an
enormous global challenge. Existing community projects running elsewhere were seen as potentially helpful, such as home energy/carbon audits
to encourage reduced CO_2 emissions through shifts in behaviour and the
uptake of energy efficiency measures including improved insulation.

The HadLOW CARBON Community became increasingly involved
in such assessments. However, a frequent comment was: 'What good can
come out of our efforts when the world is so threatened?' This uncertainty
about action to cope with climate change has been studied by Grothmann
and Patt (2005, p. 209), who note that: 'perceived adaptive capacity has
largely been neglected in the previous literature on adaptation to climate
change'. These authors add that socio-cognitive factors appear to be more
important than objective socio-economic issues for their case studies; that
is, the ability of people to think and re-orientate their views is important.

More recent research on community-based adaptation to climate change
(Ebi and Semenza, 2008) discusses the relevance of the three most commonly distinguished forms of social capital, namely 'bonding', 'linking'
and 'bridging'. Bonding social capital is considered to be the 'social glue'
which holds together homogeneous groupings, such as religious, cultural,
professional, racial or ethnic groups. As Ebi and Semenza state: 'bonding
social capital is necessary but not sufficient to address the threats from
climate change' (p. 502). Linking social capital is thought to be more
important in this context because it 'connects people at different levels of
power, such as community members and government officials' (p. 503).

Linking social capital has certainly been important for the HadLOW
CARBON Community in securing financial support from local councils,
for example from Kent County Council for the Finite Exhibition. Thus,
as the HadLOW CARBON Community enjoys more project success,
the value of income via linking social capital increases due to better

development of trust with local government and the associated likelihood of support.

Bridging social capital 'arises from connecting socially heterogeneous groups and can provide a host of benefits to community groups. Different societal groups vary in skills and talents and can generate new strategies for addressing risks' (Ebi and Semenza, 2008, pp. 502–3). It is argued that the HadLOW CARBON Community has benefited most from bridging social capital – the power of cross linkages between the college and village, as illustrated in the above two projects. The success of the Grower Group has so far depended upon novel linkages between people with vastly different reasons for being in Hadlow: college teachers and students are there to teach and learn, respectively and whilst some village residents also work locally, including at the college itself, most want simply to live in Hadlow and enjoy leisure and a good quality of life. The most powerful link between these groups has been horticulture and countryside skills – college stakeholders with a professional commitment and village residents with a private interest. Thus, a joint focus on the growing of vegetables for food has formed the bridge, with countryside skills such as fencing acting in support. This project continues though the pace of development varies, for example with availability of students limited over each summer. Currently a recruitment drive is underway to widen village participation.

For the Finite Art Exhibition, art as a medium for expression has also proved to be a powerful bridge between villagers and the college. The College Garden Design Centre which hosted the exhibition is devoted to the use of art as media in the visualization of options for garden development. Similarly, the Finite Art Exhibition sought to utilize art media in helping the public visualize and understand climate change. Thus, the use of art as media for expression has proved to be an effective bridge. Such collaboration is planned to continue: other versions of the Finite Exhibition have recently been shown at venues in the county and the college, and HadLOW CARBON Community is discussing how art can be better developed as media for sustainability education.

The developments of the HadLOW CARBON Community can thus be described as a paradigm shift – that is, a transformation that is driven by agents of change (climate change in this case). The development of bridging social capital by the HadLOW CARBON Community initiatives described for Hadlow is seen as an ongoing process, reaching out to all sectors of the community irrespective of gender, age, ability, culture and socio-economic background. Other low carbon community initiatives without access to a nearby college can develop similar links with local companies, schools, councils, farms and other private land owners. The latter are now being linked to Grower Groups seeking suitable space to grow food, via the new 'Landshare' programme (Landshare, 2009).

The Outcomes of the HadLOW CARBON Community

So far, the two projects described in this chapter are unique to the HadLOW CARBON Community but, in common with other low carbon community initiatives, additional HadLOW initiatives are underway, such as carbon audits and reduced residential emissions, film shows, car sharing and the development of new cycle tracks. Collectively, what tangible outcomes can be charted?

Mitigation v. adaptation

Several recent research studies have attempted to shed light on the interplay between climate change mitigation and adaptation in sustainable communities (Callaway, 2004; Dessai et al., 2005; Grothmann and Patt, 2005; Ebi and Semenza, 2008). Callaway (2004) for example emphasizes the need to:

> debate the idea that the global marginal benefits of mitigation cannot be compared with the local marginal benefits of adaptation. In fact we can do this, by using information about the degree of 'substitutability' between emissions reductions and adaptation in reducing local damages to translate local marginal adaptation benefits into their local emissions reduction benefit equivalent. (p. 273)

The HadLOW CARBON Community Grower Group plot will allow local people to adapt to climate change by means of techniques to improve the resilience of vegetable production in the face of extreme weather events, for example by using enhanced organic matter (Cook et al., 2006) and raised beds (Wright, 2008) to improve drought tolerance. In so doing, Grower Group members are substituting local vegetable production for imported equivalents and associated emissions.

Green exercise and community identity

There has been much recent investigation of green exercise in local communities, especially in rural areas like Hadlow (see a review by Pretty et al., 2007). Such research identifies the additional benefits of regular exercise in a natural setting, both in terms of physical and mental well being. A study of visitors to parks and forests in Switzerland indicated that 98.4 per cent had subjectively enjoyed benefits in terms of reduced stress levels and the incidence of headaches (Hansmann et al., 2007). Likewise, Pretty et al. (2005, 2007) undertook case studies which showed a significant improvement in self-esteem and mood as a result of activities such as boating, walking, conservation, horse riding and fishing. It is suggested here that Grower Group activities are also likely to have similar benefits and are worthy of study.

There is also potential for investigation of the psychological benefits of being part of a low carbon community initiative such as the HadLOW CARBON Community. Rudd (2000) indicates clear mental advantages to membership of such a community; that is, enhanced self-esteem and general feelings of well being. Anecdotally, discussions with HadLOW Grower Group members have highlighted the personal benefits of involvement, though this has not been formally studied.

CONCLUSIONS

There is argued here to be a large potential for positive benefits from low carbon community initiatives such as the HadLOW CARBON Community. These include enhanced adaptation to and mitigation against climate change but also significant improvements in the physical and mental well being of participants. Thus low carbon community initiatives are proposed as powerful structures that operate most effectively in terms of the development of 'bridging social capital', which is suggested as the vision for the future of adaptation to and mitigation against climate change at the local level. The HadLOW CARBON Community is argued to be a working example of bridging social capital, where Hadlow village has 'bridged' across differences to collaborate with Hadlow College. This is illustrated by the collaboration required for the Grower Group, and also that for joint art projects such as the River Knit display at the Finite Exhibition.

These collaborative and dynamic relationships for the development of bridging social capital are needed elsewhere, perhaps linking to other stakeholders such as small and medium-sized enterprises, local councils and non-governmental organizations. Low carbon community initiatives also need to network and exchange best practice. Thus it is concluded that low carbon community initiatives which outreach to and collaborate with other stakeholders can be a powerful force against climate change and a positive option for a healthier and more sustainable future for us all.

ACKNOWLEDGEMENTS

The authors would like to thank many village residents and members of the HadLOW CARBON Community for their support, and especially

Stephanie Ingham and Nicola Canham. There are also numerous Hadlow College staff to acknowledge but specifically we would like to thank Pam Worrall for helpful comments on the manuscript and the Director of Finance, Mark Lumsdon-Taylor, and Sue Brimlow in the Business and Community Development Unit, for their enthusiasm and commitment.

NOTE

1. In the UK, green belt areas are designated primarily for agriculture, forestry and rural outdoor leisure and are usually planned to resist encroachment from urban development.

REFERENCES

Callaway, J.M. (2004), 'Adaptation benefits and costs: are they important in the global policy picture and how can we estimate them?', *Global Environmental Change*, **14**, 273–82.

Cook, H.F., G.S.B. Valdes and H.C. Lee (2006), 'Mulch effects on rainfall interception, soil physical characteristics and temperature under *Zea mays*', L. *Soil & Tillage Research*, **91**, 227–35.

Daniels, S. and G.H. Enfield (2009), 'Narratives of climate change: introduction', *Journal of Historical Geography*, **35** (2), 215–22.

Dessai, S., X. Lu and J.S. Risbey (2005), 'On the role of climate scenarios for adaptation planning', *Global Environmental Change*, **15**, 87–97.

Ebi, K.L. and J.C. Semenza (2008), 'Community-based adaptation to the health impacts of climate change', *American Journal of Preventative Medicine*, **35** (5), 501–7.

Gablik, S. (1997), *Conversations Before the End of Time*, London: Thames & Hudson.

Grothmann, T. and A. Patt (2005), 'Adaptive capacity and human cognition: the process of individual adaptation to climate change', *Global Environmental Change*, **15**, 199–213.

Hansmann, R., S.R. Hug and K. Seeland (2007), 'Restoration and stress relief through physical activities in forests and parks', *Urban Forestry and Urban Greening*, **6**, 213–25.

Howe, K. (2009), personal communication with Tunbridge Wells Transition Town Group, 26 May.

Landshare (2009), 'Linking people who want to grow their own food to space where they can grow it', accessed at www.landshare.net.

Pears, P. (2004), *Growing Fruit and Vegetables on a Bed System the Organic Way*, Tunbridge Wells: Search Press.

Pretty, J., J. Peacock, M. Sellens and M. Griffin (2005), 'The mental and physical health outcomes of green exercise', *International Journal of Environmental Health Research*, **15** (5), 319–37.

Pretty, J., J. Peacock, R. Hine, M. Sellens, N. South and M. Griffin (2007), 'Green exercise in the UK countryside: effects on health and psychological well-being, and implications for policy and planning', *Journal of Environmental Planning and Management*, **50** (2), 211–31.

Rudd, M.A. (2000), 'Live long and prosper: collective action, social capital and social vision', *Ecological Economics*, **34** (234), 131–44.

Salinger, M.J., C.J. Stigter and H.P. Das (2000), 'Agrometeorological adaptation strategies to increasing variability and climate change', *Agricultural and Forest Meteorology*, **103**, 167–84.

The Eden Project (2009), www.edenproject.com, The Eden Project, Bodelva, Cornwall, PL24 2SG, UK.

Wright, J. (2008), *Sustainable Agriculture and Food Security in an Era of Oil Scarcity: Lessons from Cuba*, London: Earthscan.

15. Empowering farmers to react and to act: from an anti-golf course pressure group to a community-based farmers' cooperative

Mario Cardona

MANIKATA: FEELING THE PLACE

I arrived in Manikata when it was almost dusk . . . The silence was so thick that it could be almost felt as it added to the peacefulness of the surroundings. As one looked down towards the sea one could make out the outlines of the dwellings where the inhabitants of this lovely hamlet live their life full of toils, joys, loves and pains as is normal with most humans when in their homes.

But these people have a gnawing fear, real fear. Their term of tenure for the land they cultivate there is being terminated. A small note and hey presto your permit to work here on this land, where you were born as were your fathers and grandfathers before you, is withdrawn. We are the government. We know what is best for you. Think what you like, but do what We tell you, because we are democratic. Democracy oozes out from all Our political pores.

The silence which induced these reflections was broken by the hum of a sizeable throng of people that started coming out of the local parish church bidding each other good night. They went there to pray to Our Lady in this month of October. Their belief is still strong and deep. The church they came out of was erected by their donations . . . We read that one's faith is as deep as one's hand can go into one's pocket. These generous citizens of Manikata must have very deep, strong faith.[1]

The letter above was written at a crucial turning point in the history of the small rural community of Manikata. Manikata is a small modern village in the northwest of the island of Malta, and is home to about 500 people, many of them farmers. It is built on the slope overlooking the agricultural district of Għajn Tuffieħa, an irrigated valley bed that runs from St Paul's Bay to Golden Bay. Towards the end of June 2005, tenant farmers on state owned land started receiving letters signed by the President of the Republic stating that they were to be evicted from their fields and, in the case of two families even their homes, in order to make way for a golf course that was to surround the western side of the village and cushion it off from the nearby coast. At the same time, news started reeling in that the Malta Transport Authority was going to build a motorway across the valley to the southeast, cutting across the fields, including a tunnel that was to be dug metres below the village core and emerging to the north, cutting across the other valley that separates Manikata from the nearby town of Mellieħa. The news hit the local community by storm. The sleepy rural village that very few people in Malta knew about was going to be obliterated. This chapter considers how the local community dealt with this awkward situation and with the way it managed to use the skills learned during its campaigning for a positive end.

AN EMPOWERING CAMPAIGN

The 'Achieving Better Community Development' framework developed by the Scottish Community Development Centre includes four central dimensions of community empowerment, namely: personal empowerment, positive action, community organization, and participation and influence. According to this framework these dimensions lie at the heart of 'liveable, equitable and sustainable communities' (Ledwith, 2005, pp. 80–81). In varying degrees, these four dimensions characterized a two-year campaign in Manikata that matured from a state of inertia and fear to one marked by self-fulfilment on an individual and collective level, the learning of new skills, knowledge and attitudes, and the desire to become directly involved in the shaping of the community's future through the creation of a community-based project.

Rumours were rife in Manikata, and the usually calm atmosphere of the village was being agitated by strong undercurrents. On Sunday 17 July 2005, at the end of an anti-golf course protest march organized by Alternattiva Demokratika[2] in Manikata, two part-time farmers decided to take action to save the village. They asked me if I would be willing to chair a community meeting they were going to summon for the following Sunday at the parish hall, put at their disposal by the parish priest. I accepted with some trepidation, but also excitement. The two farmers distributed flyers in all the households of Manikata and the next Sunday the meeting was held, early in the afternoon. This in itself was a watershed in the history of the community. The only time in the recent history of Manikata that something similar had happened was between 1902 and 1905, when the British Admiralty took control of a number of fields in order to develop them into a Royal Marines Training Centre. No records of any reaction by local farmers with regard to this development have ever come to light. It was only the Superintendent of Public Works who tried to make sure that farmers were compensated for the produce they were going to lose once they were evicted from their fields. Perhaps no farmer would have thought of challenging the 'might of the Empire' in the early years of the twentieth century. But now the situation was different. The Empire was long gone and the state was run by Maltese politicians. The dimensions of the land that was going to be taken over by the state had no precedents. Freire (1998) states that

> Whether facing widely different challenges of the environment or the same challenge, men are not limited to a single reaction pattern. They organize themselves, choose the best response, test themselves, act, and change in the very act of responding. They do all this consciously, as one uses a tool to deal with a problem. (Freire, 1998, p. 3)

It is apparent that these two farmers felt impelled from the outset to act; simple resignation to a fate determined by central authority was out of the question. Ultimately they felt that they could engage with the current situation as subjects not as objects, turning a problem into a challenge, with potential possibilities for self affirmation, echoing Freire's argument:

> Whether or not men can perceive the epochal themes and above all, how they act upon the reality within which these themes are generated will largely determine their humanization or dehumanization, their affirmation as Subjects or their reduction as objects. For only as men grasp the themes can they intervene in reality instead of remaining mere onlookers. And only by developing a permanently critical attitude can men overcome a posture of adjustment in order to become integrated with the spirit of the time. (Freire, 1998, p. 5)

The result was the setting in motion of a very lively public debate that culminated in the drawing up of a declaration stating the reasons why the community opposed the golf course. Those present during the first public meeting also elected a committee adopting the name of Kumitat għall-Ħarsien Rurali ta' Għajn Tuffieħa.[3] All the members were full-time or part-time farmers, except for myself and a 16-year-old girl who was a member of the parish youth group which had been helping local farmers to organize their annual agricultural show. It was decided to focus on the golf course question first since we only had the official plans and reports regarding this project. When the plans for the motorway were made public, a similar process took place: the committee called for a public meeting and a declaration was drawn up clearly articulating the objections held by various members of the local community. The two documents which were drawn up were the result of an open discussion where everyone in attendance had the opportunity to voice their opinions on the impact of the proposed projects. It transpired that some people felt they would be affected in similar ways, while others expected to suffer particular adverse effects. Ultimately, reflecting as many voices as possible, the two documents emerged as a strong communal statement, based on the views of people deeply rooted, almost literally, in the land that was going to be affected by the two projects.

The next step was to decide what to do with these documents. Freire states that 'Once man perceives a challenge, he acts. The nature of that action corresponds to the nature of his understanding' (Freire, 1998, p. 44). Referring to feminist struggles in Latin America, particularly to the Mothers of the Plaza de Mayo in Argentina, Kane comments: 'paradoxically, it was precisely by playing on the traditional role of motherhood that their political struggle was so effective, leading the Mothers to adopt roles which were anything but traditional' (Kane, 2001, p. 113). Similarly, it was by emphasizing the traditional role of the farmer as the guardian of rural life with all its ramifications that the farmers on the Manikata committee came to adopt roles during the two-year struggle that were anything but traditionally associated with farmers, used as they were to let others speak for them. Initially, I was invited to chair the first community meeting and then also to chair the committee that was formed by volunteers among those present, primarily resulting from a combination of trust in my abilities as a community activist (active mostly in the parish), but also out of a lack of self confidence on the farmers' side in terms of feeling able to speak out.

According to Don Milani, an Italian priest and educator, the greatest hurdle the poor farmers in his parish had to overcome in order to become equal citizens was the acquisition of *the word* (Gesualdi, 2006). He argues that people in power have a command over the word, a hegemonic dominance over the way language is used in a sophisticated way in order to exclude those who, albeit citizens of supposedly equal status in front of the law, are actually second class citizens because they do not have the language skills, oral and written, in order to participate and impinge upon the politics of society. Writing to a friend of his who was a judge, Don Milani implores him:

> Do you want that the poor rule immediately? Do you want that they rule well? Then write a book for them, or a newspaper for them or else become an apostle among your Catholic, university graduate friends in order to set up a grandiose popular school in Florence. Not as a gift to be given to the poor, but as a debt to be settled and a gift to be received. Not to teach, but only to give the necessary technical means (that is the language) to the poor so that they can teach you the inexhaustible riches of equilibrium, of wisdom, of concreteness, of potential religiosity which God has hidden in their hearts almost to compensate them for the cultural inequality of which they are victims. (Gesualdi, 2006, pp. 44–5)

I saw that my greatest challenge was to use the fortnightly committee meetings to discuss the main issues with the farmers, help them articulate their arguments, widening their lexical register along the way, so that eventually, during the meetings and activities we would hold in our campaign, they would be able to speak for themselves rather than let me speak for them. Eventually, what followed was a long series of meetings with major stakeholders where I would usually set the discussion going, then invite the farmers to speak, enabling the interlocutor to look at the whole issue from the viewpoint of someone whose livelihood depended on the same lands that were now under threat from the proposed developments. The committee met, among others: a delegation of the Maltese Cabinet made up of three ministers and two parliamentary secretaries; the leader of the Opposition together with the shadow ministers for agriculture and the environment; the general secretaries of the three Maltese political parties; Maltese and German MEPs; non-governmental organizations (NGOs) active in the field of environmental protection; the secretaries of the two national farmers' associations; the Archbishop;[4] the Church's Commission for the Environment; the Mellieħa Local Council; and also a high level delegation from the Malta Transport Authority.

Members of the committee also attended parliamentary sessions, making sure they were noticed by the MPs of the northwestern electoral district of both parties in the House, and also took part in two national protest marches organized by Flimkien għal Ambjent Aħjar[5] against the way the government and its agencies were sanctioning unsustainable development projects in Malta and its sister island Gozo.

The committee also led an extensive media campaign, organizing press conferences on-site and keeping an ongoing correspondence in the local press.[6] It was very common during our fortnightly meetings to spend the last part of the meeting discussing letters for the press written by different members of the committee who had never written letters for publication before. Ultimately, it was the authentic voice of the farmers, and not my voice as an adult educator and community activist, which had the greatest impact. The front page of *The Sunday Times of Malta* of 21 August 2005 reported the press conference organized in Manikata by the committee during the previous day, quoting extensively the farmers' interventions during the event:

> Piju Mercieca, from Manikata, who tills 10 tumoli [approximately 12,000m²] of land, is one of the farmers who last month received a notice of termination of lease signed by President Fenech Adami. 'They are not sending all the notices at once because they do not want farmers to unite as a single front,' Mr Mercieca, a father of four, said . . . Mr Mercieca said it was ironic that the government wanted him out of his fields after having given him subsidies to

cultivate vineyards . . . 'Before the EU referendum, Dr Fenech Adami, then prime minister, had visited my fields and told me that we would be more protected once Malta became an EU member. He said Europe protected farmers, and that my livelihood would be guaranteed if I invested in more modern systems. It's really ironic that now, as President, he has had to sign my notice of termination of the land lease,' Mr Mercieca said. (*The Sunday Times of Malta*, 21.08.2005, p. 1)

The excerpt above highlights the way arguments refined during the committee meetings were put to the media by the farmers themselves with vigour and forcefulness. In terms of Freire's discourse this is an indication of the way the farmers managed to 'perceive and respond to suggestions and questions arising in their context, and increase their capacity to enter into dialogue not only with other men but with their world . . . by refusing to transfer responsibility; by rejecting passive positions; by soundness of argumentation; by the practice of dialogue rather than polemics' (Freire, 1998, pp. 17, 18).

THE AFTERMATH

The campaign that started on 17 July 2005 was a long and hectic one. Then, after a period of silence and no communication from official quarters, the committee received an official letter from the Malta Transport Authority, on 26 October 2006, to inform them that the proposal for the Xemxija Bypass via Manikata had been withdrawn. It is difficult to express the feeling in such moments. When we started out, many people expressed scepticism with regard to the efforts of the farmers; in effect 'What can you do against the government? The land ultimately belongs to it and it can do anything with it. It is useless trying.'

This is where we took off from. But it was fulfilling for me, an adult educator involved in community development, to see the farmers on the committee taking microphones, megaphones, pens, laptops or keyboards in hand, and publicly denounce the great injustice that they felt was going to be meted out to them; show their face, put their name to every letter to the editor, public statement or press release. They had initially hired an adult educator to speak up for them. But ultimately they spoke for themselves, both as individuals and as a collective entity. They had believed that they could do something to enable government to recognize the negative impacts of its development plans in their locality. They also came to believe that collective action, based on mutual trust, was the way forward for farmers. Traditionally the mindset of Maltese farmers has tended (unofficially) to be associated with individualism and distrust. Borrowing

Mayo's terminology, in this situation it seems that a two-year campaign had enabled the farmers to *decolonize* their minds (Mayo, 2007, p. 110).

The receipt of Malta Transport Authority's letter regarding the shelving of the Manikata motorway project spurred the committee on to face the remaining issue: the golf course. We noticed that after the widespread popular resistance to the project, fomented by our committee's lobbying as well as by the equally resilient campaign led by various environment and heritage organizations, the government appeared to have put the project on hold, mulling over a possible withdrawal, and perhaps waiting for the 'right' moment to do so, in order to limit political damage.

Our committee sent several letters to various ministries but failed to receive a definite answer. We thus kept applying pressure with letters in the media and participation on radio talk shows. In the meantime, unofficial sources started sending messages informing the committee that the project was going to be shelved and that it was just a question of time. We were impatiently waiting for the official news to break.

The campaign coincided with my own studies at the University of Malta, where I was exposed to the writings of Paulo Freire. Almost unavoidably, my involvement with the Committee for the Safeguard of Rural Life at Għajn Tuffieħa was very much influenced by the Brazilian pedagogist's writings. Ledwith (2005) argues that 'Freire achieves theoretical coherence because his work unites a philosophy of hope with a pedagogy of liberation. The basic belief underpinning this is that human beings are subjects, able to think and reflect for themselves, and in doing so transcend and recreate their world' (p. 96). In a clearly Freirean frame of mind Ledwith comments on the process of community development which arises from problems but leads to positive action, stating that,

> The hopelessness that gives rise to anger or apathy becomes a more dignified and determined hopefulness. A world held in common is one in which we are able to reach across all aspects of difference to act together on issues that are wrong. In these ways, an altered way of 'knowing' the world (epistemology) results in changed ways of 'being' in the world (ontology). (Ledwith, 2005, p. 2)

The point at which the members of the committee felt they had been 'strengthened' enough to pass from a defensive to a proactive stance, and to develop a proposal of their own for their fields, arrived in late 2006. The proposal focused on the Razzett tal-Qasam, a historic farmstead dating to at least the eighteenth century, lying in a derelict state on the outskirts of Manikata.

Until the immediate post-war period it housed a number of families in its various rooms and caves. These families then moved out, going on to live in modern houses in the new developing urban hub of Manikata, about half a kilometre uphill. The last remaining family moved out in about 1980 and the tenancy became vacant. Soon afterwards the farmstead was subjected to vandalism and left severely damaged. Some of the surrounding fields were also abandoned because, lying on the sloping cliff face, there is insufficient soil on the surface and the farmers preferred to work other, more fertile fields further down on the valley bed. Both the farmstead and the surrounding fields were designated to become part of the golf course development.

The committee proposed that we ask the government to lease the farmstead to us in order to restore it and turn it into an agro-tourism centre, where farmers would be able to process products and sell them to visitors. Tenant farmers who had the lease of the surrounding unutilized fields would then be encouraged to convert them into olive groves. It was proposed that space in the restored farmstead could be used to set up an olive press and market our own olive oil. In this way the abandoned fields could be recovered back to agricultural use, landscaping of the area enhanced and a farmstead of historical and architectural value restored. All this would be obtained through a sustainable project that would be economically viable and hopefully profitable for the farming community. It was intended that such a project would also nurture a culture of cooperation between farmers, something which goes against the grain of rural culture in Malta, where traditionally farmers have tended to work in isolation.

This proposal needed the backing of government, the owner of the abandoned farmstead. The unofficial messages about a possible permanent shelving of the golf course project seemed to have made possible a new kind of relationship with the authorities. Speaking about popular movements, Marleny Blanco states that,

> On this unfolding path of opportunity, we come across new forms of relationship between government and society. It has to be said that it is not due to the kindness of governments that these new spaces emerge. They are the result of the popular movements' struggle, not only of protest and pressure but also their ability to make positive, concrete proposals for change. (Marleny Blanco, 2001, p. 168)

It was a result of the two-year campaign against state sponsored projects that led the community to operationalize a project where the farmers would emerge as the prime movers. It was as if the golf course and the motorway threats had awakened the local farming community to envision new possibilities for the locality and the community. This point is very well illustrated by Davidson when he argues, 'I would . . . question the liberal assertion, rapidly becoming received wisdom, that positive change in rural communities can only arise from within and that external influences are always destructive and disempowering. In many cases it is the "irritant" from outside the community that proves to be the catalyst for change' (Davidson, 2002, p. 22).

We sought the backing of the Mellieħa Local Council and together we proposed the project to the minister in charge of the Lands Department, who informed us that he would be unable to provide any support because the golf course project was still being discussed. We knew we had to wait.

In order to shorten that time as much as possible, we initiated contact with Maltese officials in Brussels, sending informal messages to government that if the definite decision on the golf course would be long in coming, we would take the issue to the EU level. On Saturday 12 May 2007 the Maltese Prime Minister held a press conference announcing that the golf course project had been shelved and that the land would instead become part of a national nature and history park: the Park tal-Majjistral.[7] On the same day the committee was informed in an email from the Prime Minister that all the farmers involved would be notified shortly to renew the tenancy on their lands. We immediately wrote back expressing our feeling that a national park would be much more compatible with the economic, environmental and cultural features of the area, and that we hoped to be consulted on the way the park would be set up and managed. The postscript to our email communication read:

> If there is one good thing that came out of this affair, it is the sense of solidarity among the farmers active within our committee. The sense of solidarity has led our committee to embark upon an agro-tourism project . . . which among other things, involves the restoration of an extensive abandoned farmstead . . . The mutual trust built among farmers during this campaign will be channelled into a positive project.[8]

The Prime Minister came to visit Manikata farmers soon after and we explained the project to him. Subsequently, the Minister for Rural Affairs and the Environment, the Parliamentary Secretary for Agriculture and a number of directors from the Agriculture Department met with the committee to discuss the project. We visited the site together and on the strength of this were invited to prepare and submit a full project proposal. By March 2008 a management agreement for the site (renewable every five years) had been secured.

THE KOPERATTIVA RURALI MANIKATA LTD

Following the meeting with the Prime Minister we had lengthy discussions among ourselves, and also with a number of experts in community development and representatives from the Maltese cooperative movement, in order to develop a workable strategy for the farmstead project. A public meeting was called in order to report back to the community regarding the committee's progress over the previous two years, and to propose the establishment of a new cooperative to replace the existing committee and take responsibility for managing the Razzett tal-Qasam agro-tourism project. The Maltese cooperative movement is traditionally linked to agriculture, though lately it has started to infiltrate other business areas. A cooperative model was chosen in order to consolidate and build upon the knowledge, skills and attitudes learned during our two year campaign, marked as it was by a spirit of cooperation and involvement together with commitment towards the local agricultural community. Delia (2006) outlines the advantages of the cooperative organizational model thus:

> The merits of the co-operative model of organisation may be considered in terms of the role institutions play over time in the social and economic development of a region . . . The merits of a co-operative may also be evaluated in relation to its contribution to personal fulfilment in the context of specified ethical goals. Personal and social values such as solidarity and wide participation in economic gains have to be considered in tandem with economic efficiency and sustainable economic growth. In this context, efficiency and co-operation constitute one element in the wider debate on the role of production and a fair income distribution over time in a society. (Delia, 2006, p. 59)

The public meeting was held on 14 June 2007. The members of the community present discussed the two-year campaign and thanked the farmers of the committee for their time and energy. A cooperative proposal was subsequently put forward. Most of those present considered the proposal a good idea; however, not all were sure about becoming involved in a longer-term project. Eight people agreed to join in, five of them members of the original committee. This was enough to begin the process in respect of the Maltese cooperatives law, which sets the minimum number of members at four. These eight people formed a steering committee who, with the help of people from other cooperatives and from the Central Cooperatives Fund, drafted a statute and also contacted others who they thought might be interested to join in the cooperative project. Within days, the new cooperative had 14 members. An application was filed and an appointment secured with the Cooperatives' Board to discuss our application. Two farmers from the steering committee met the Board to discuss the proposal. The Board were convinced by the sustainability and feasibility of the new cooperative and subsequently gave consent for the establishment of the Koperattiva Rurali Manikata (KRM) Ltd.

THE PROJECT IN ITS FIRST YEAR

Once the first management committee had been elected, work began to find practicable ways to set the farmstead project in motion. From the outset, we wanted to put the project into a wider perspective that would focus not only the restoration of the abandoned Razzett tal-Qasam, but also include: the recovery of abandoned fields around the farmstead back to agricultural use through the planting of olive trees; the eventual installation in the restored farmstead of an olive press for the production of olive oil; a kitchen for the production of jams and preserves that would recover traditional local recipes by finding a suitable market for their sale; and an evaluation of the archaeological and historical buildings/remains along the fields surrounding the farmstead. The Razzett tal-Qasam project would thus have aesthetic, social, economic, cultural and environmental dimensions at its core. What seemed to bring the project together as a whole were plans for a Rural Heritage Trail, starting from a small animal farm, proceeding through the ruins of the Razzett tal-Qasam farmstead and meandering through the fields where visitors would be encouraged to appreciate and explore several points of interest, including a 1935 British military beach post, a Roman tomb later used as a Second World War air raid shelter and eventually as an apiary, historic corbelled stone huts, indigenous trees and shrubs, a vineyard, a market garden, and the evolving olive groves. At the end of the trail visitors would be given traditional bread, spread with local tomato paste and olive oil, some fresh fruit, water and wine from the Manikata vineyards. Members of the cooperative would start producing typical local jams and vegetable preserves such as pickled onions, sun-dried tomatoes, olives, honey and carob syrup. These would be sold to visitors to the trail. It was hoped that the Rural Heritage Trail would eventually enable visitors to gain fresh insights about the countryside and get a feeling of what a rural village like Manikata really offers. In other words, to experience the fact that 'the rural was never timeless and tranquil, but a place of work and living as well as rest and retreat' (Davidson, 2002, p. 23).

Kane argues that 'However sympathetic the aim, attempts to eulogise without problematising indigenous cultures ultimately do the cause of indigenous people a disservice' (Kane, 2001, p. 126). In fact, the establishment of the trail enabled the members of the cooperative to question old practices such as that of using abandoned fields as scrap yards or as deposits for unwanted car batteries, broken furniture or even scrapped cars; or the practice of giving a free hand to hunters who literally took over the farmers' abandoned fields around the farmstead, turning the military beach post into a hunting hide and the Roman tomb into a toilet. We also questioned the culture of considering fields only in terms of economic return without any consideration for the aesthetic dimension, which had clearly contributed to indifference towards collapsed rubble walls and corbelled stone huts. Through the setting up of the Rural Heritage Trail these practices started to be reversed. Working together, the members of the cooperative cleaned the trail and the area of the farmstead from tonnes of scrap material; sponsors were found for the planting of 700 olive trees in the formerly abandoned fields; sponsors were also found for the restoration of two dilapidated corbelled stone huts; the beach post and the Roman tomb were cleaned up and replicas of traditional clay honey pots reinstated in the apiary.

Funds were secured from the EU Leonardo Programme, enabling the cooperative to send two members on a two-week hands-on training course in Sicily, where they visited various agro-tourism centres scattered along extensive olive groves in the area around Syracuse. Another member of the cooperative's committee helped a member of the University of Malta's Agricultural Institute to bring over from Bari two experts in olive tree cultivation and pruning to conduct a two-day training programme. Volunteers from the cooperative cleaned the Razzett tal-Qasam from the invading prickly pear trees and an architect was commissioned to draw up a survey of the area and file an application with the Malta Environment and Planning Authority for its restoration and renovation. In its first year of operation the heritage trail attracted more than 3000 visitors. Profits were invested in the clean-up and surveying of the Razzett tal-Qasam, where the installation of an olive press is planned within three years, when it has been projected that the first sustainable olive crop will be produced. With the help of the Institute of Tourism Studies, we organized a course in the production of typical local food and have found a profitable market for local food products which are becoming very popular with local and foreign visitors to our trail, with additional profits for the farmers.

The feedback received from students, families and tourists so far has been positive. It indicates that visitors are appreciating the different aspects of rural life, the environmental landscape, the sounds, the colours, the smells of the different wild plants and shrubs, the contact with farm animals and the taste of typical local food – taken together, a real opportunity to experience a different rhythm of life from that of Malta's urban centre.[9] It is emblematic that most teachers accompanying students to the trail tell our programme animator how they usually relish the bread with tomato paste we give them at the end of the trail, not because there is anything special about it but simply because eating that kind of bread in a rural setting in the shade of carob and pine trees is something 'different'. In Davidson's words, 'It is in the rural that the traditions of a nation or a region are often assumed to be most clearly embodied, and that within these traditions there exists a rural or native wisdom, free of the artifice and superficial sophistication of the metropolis with its inevitable processes of commoditization' (Davidson, 2002, p. 20).

THE TASKS AHEAD . . .

The setting up of Koperattiva Rurali Manikata was the result of the way in which all the members of the Committee for the Safeguard of Rural Life in Għajn Tuffieħa transformed their campaign into a learning experience – a milestone in the history of community development in Manikata. Speaking about the role of cooperatives in Malta, Delia argues that,

> In future, the co-op has to become an instrument that enables fair competition for both producer and/or consumers. This can be achieved if life and vision of co-operatives arise from 'below', from a feeling among its potential members that they can support one another, think collectively, and act accordingly in order to compete in free markets. Short of this multiple vision, a unit claiming to be a co-operative will be assuming the legal structure of a co-op but will fail to live up to the co-operative spirit. (Delia, 2006, p. 89)

The Razzett tal-Qasam – which to a large extent encapsulates the locality's history and rural origins – is slowly being restored by a number of farmers working together to bring this farmstead back to life. Abandoned fields are being restored back to agricultural use through the planting of a number of Maltese and European olive tree varieties. Locally produced food, typical to the area, is being reappraised by the local community, and highly appreciated by visitors. The communication and organizational skills learned during a two-year campaign to save our village are being put to use in piloting our community project. The Razzett tal-Qasam, slowly but steadily rising like a phoenix, embodies a locally based project where economic sustainability, environmental protection and community interests are integrated. The corbelled stone hut, formerly lying partly in ruins in a forgotten corner in the fields, today has acquired a new significance; a humble but self-respecting symbol of the social, cultural and economic wealth of our rural district.

NOTES

1. From a letter to the editor written by Joseph Farrugia, secretary, Progressive Farmers' Union, after meeting the Kumitat għall-Ħarsien Rurali ta' Għajn Tuffieħa (Committee for the safeguard of Rural Life at Għajn Tuffieħa). *The Times of Malta*, 31.10.2005.

2. The Maltese Green Party. It has so far not been able to secure a seat in parliament. At the time it was represented on various local councils in Malta and Gozo and has been crucial to put environmental issues on the national political agenda.
3. Committee for the Safeguard of Rural Life at Għajn Tuffieħa.
4. The parish priest accompanied the committee in this meeting.
5. Together for a Better Environment, environmental NGO.
6. See for example, *The Sunday Times of Malta*: 21.08.2005, 23.10.2005, 05.03.2006, 26.03.2006. *The Times of Malta*: 15.09.2005, 26.09.2005, 28.10.2005, 31.10.2005, 09.11.2005, 21.11.2005, 28.11.2005. *L-Orizzont*: 27.08.2005, 02.09.2005, 19.09.2005. *It-Torċa*: 21.08.2005, 28.08.2005, 20.11.2005, 22.01.2006, 26.04.2006. *Il-Mument*: 07.08.2005. *Malta Today*: 21.08.2005, 04.12.2005. *The Malta Independent on Sunday*: 21.08.2005, 05.03.2006.
7. The North Eastern Park.
8. From an email to the Prime Minister, 12.05.2007.
9. Malta is really one big city state, with an extensive urban sprawl pivoting on the Grand Harbour area. Its open rural spaces are mostly found around the coast from the south-east harbour of Marsaxlokk to the northwestern cliffs around Mellieħa. Manikata and its rural surroundings are found in the northwestern region.

REFERENCES

Davidson, I. (2002), 'Rural society, social change and continuing education: from Wild Wales to the Aga Saga', in F. Gray (ed.), *Landscapes of Learning: Lifelong Learning in Rural Communities*, Leicester: NIACE, pp. 19–31.

Delia, E.P. (2006), 'Economic efficiency, solidarity, and the co-operative model: lessons for Maltese co-ops', in E.P. Delia (ed.), *Reconsidering Co-operatives: Lessons for Maltese Co-ops*, Valetta: APS Bank, pp. 55–91.

Freire, P. (1998), *Education for Critical Consciousness*, New York: Continuum.

Gesualdi, M. (ed.) (2006), *Lettere di Don Lorenzo Milani Priore di Barbiana*, Turin, Italy: Oscar Saggi Mondadori.

Kane, L. (2001), *Popular Education and Social Change in Latin America*, London: Latin American Bureau.

Ledwith, M. (2005), *Community Development. A Critical Approach*, Bristol: The Policy Press.

Marleny Blanco, E. (2001), 'Review of Chapter 6', in L. Kane, *Popular Education and Social Change in Latin America*, London: Latin American Bureau, pp. 166–9.

Mayo, P. (2007), '10th anniversary of Paulo Freire's death. On whose side are we when we teach and act?', *Adult Education and Development, DVV International*, **1** (69), 105–10.

Milani, L. (1997), *Esperienze pastorali*, Florence, Italy: Libreria Editrice Fiorentina.

Epilogue: retrofitting buildings viewed as a civil engineering project – just do it

Michael Kelly

For the first time in history, the world has been interconnected by a global wide band-width communication system. What has come to the fore in popular and public discourse? It is a litany of Malthusian problems confronting the planet, from overpopulation, through resource depletion, environmental degradation, climate change to the down-sides both of poverty and of affluence, and financial chaos. Furthermore, sensitive sensing techniques provide daily evidence of the way in which humankind's behaviour is perturbing the ecosystems of the planet. If we imagine someone in 2050 writing the history of this time, when we have begun to articulate the challenges in front of us, how will they judge the nature and scale of our responses?

We have had the benefit of the Stern report on the economics of climate change for three years now. The discourse in international government circles has been based almost entirely on economics. Even the otherwise excellent UK Government's recent national strategy for climate and energy, *The UK Low Carbon Transition Plan*, mentions engineering only eight times, all in passing, and the analytical supplement has no mention at all! Until we start to think of the retrofit of the existing building stock and the renewal of the national infrastructure of transport, energy, waste and water as a single integrated and complex civil engineering project to be delivered over 40 years, we will run the risk of falling far short of responding on the scale that will be necessary to reach a climate resilient, energy efficient and sustainable UK by 2050. The carbon emission reduction targets serve as a convenient, but not a totally congruent, proxy for energy efficiency and sustainable consumption.

If we focus on existing buildings, the source of 45 per cent of today's total carbon emissions in the UK (with 27 per cent of all emissions coming from our homes), and recognize that 87 per cent of these buildings will still be functioning as about 70 per cent of the building stock in 2050, then we see one big aspect of the civil engineering project: a major retrofit project. There are 22 million homes and 5 million non-domestic buildings (offices, retail outlets, warehouses, hotels, hospitals, etc.). We know from a handful of examples that £100,000 used in the makeover of semidetached homes (31 per cent of our stock) can produce a 60 per cent reduction in carbon emissions. One example indicates that only 15 per cent of the cost is associated with the extra expense of specifically low carbon options, while most of the cost is associated with a routine makeover which is necessary to get at the systems for insulation, water reduction, more efficient boilers, and so on. If we could generate economies of scale, and perhaps halve these costs, we are still talking about a £300 billion part of an overall £2 trillion investment on the renewal of our building stock. If we imagine that we will be able to come in twice between now and 2050 with a whole-house intervention to take each house initially to a 60 per cent reduction and later to the 80 per cent reduction in carbon emissions, we will need soon to be scaling to 1 million interventions a year to be on track. With 68 per cent of homes in private ownership and occupation, there is a further complication as when some interventions are most likely done far more effectively and cheaply at a communal level, such as some heat pumps or the anaerobic digestion of waste. How can this be arranged?

One of the biggest hurdles towards progress at scale in terms of retrofitting buildings is the total absence of a retrofit market (there is no entry in the Yellow Pages) and the totally balkanized structure of the existing renovation market. There is no dominant member of the sector to take a lead, as might BT, BP or Tesco in their own sectors of activity. Many of the small players are keen to get involved, but are fearful that free-riders could undercut them with inferior but cheaper work in the transition period and threaten their futures. This does seem a case for Government leadership in regulating for ever-improving standards of work, that is, an escalator of virtue, on to which the entire sector can step.

In the end, all aspirations of central government are delivered at the local level by communities. In the case of a renewal of buildings and the national infrastructure, the local element will be essential. After all, it was the likes of Joseph Chamberlain as mayor of Birmingham who oversaw the establishment of the civic infrastructure we have enjoyed for the last century and more. We will need a new generation of strong and effective civil leaders. My own contribution so far is to bid for funds so that I might produce robust trajectories and route-maps whereby urban conurbations such as Cambridge or Greater Manchester might get from here to the 2050 targets in a sensible manner. At present there is no guidance that would allow local officials to defer an action with the good knowledge that something better is in the pipeline. Suppose for example we were to delay any action on (say) Victorian terraces and townhouses for a decade: we could in the interim develop a clip-on external cladding that would be mass produced off-site and cut to precision-fit houses street by street, rather in the way we changed from town gas to North Sea gas 40 years ago. Any road map now will be inaccurate in accounting for the future, but an initial road map will be better than none, leaving local officials otherwise having to choose interventions without discrimination.

The role of R&D simply cannot be ignored. If we consider the application to houses between 1990 and 2005 of the following basket of measures, namely loft insulation to three inches or more installed, 60 per cent of rooms by volume draught proofed, 60 per cent of windows by area double glazed and cavity wall insulation installed where appropriate, then the fraction of the housing stock saving energy through these measures rose from 35 per cent to 65 per cent. At the present rate, these measures will be exhausted within 5–7 years, and 10–12 years for deep insulation and 100 per cent double glazing and draught proofing. Higher standards will be needed for thermal insulation, and these will be based on new materials, new methods of installation, and better controls on the use of energy. What is even more stark is that during the 1990–2005 period, net carbon emissions fell by 4 per cent: this was a figure that netted off the 10 per cent increase in the number of houses, and the rise in the use of energy for home entertainment and IT systems, against more substantial energy savings from the measures to improve the fabric of the buildings. The 2020 targets require us to reduce carbon emissions from the domestic sector in the period 2005–2020 at six times the net rate achieved between 1990 and 2005, and we are already 30 per cent through this latter period! Indeed if the higher and further education sectors were tasked (or better volunteered) to get their own estates to the 2050 standards by 2035, we as the public should provide them with support so that they can show us the way. In the process they will research, develop and demonstrate new technologies on their own estate: collectively the sectors are big enough to engage the construction products sector in scaling up production to that required for the nation.

Rather than be paralysed by ever more analysis of the triple challenges, there is a strong case to be made at the local level to just get on and do something about it. First, imagine a number of local authorities, working out exactly what they would do with the order of £1 billion to improve their locality in terms of energy efficiency, climate resilience and sustainability of consumption. How would the spend divide between buildings and (transport, energy and waste) infrastructure to get the deepest cuts in emissions? What is the appropriate balance of investment between mitigation and adaptation to future climates? What should be deferred so as to get much more effective action a decade hence, and what is left that should be started now?

Secondly, how could we arrange for an acceptable return on a long-term investment – energy savings, council taxes, communal transport income ...? If there are satisfactory answers to both questions, and the second will be harder at present, then we could approach the pension funds. The people joining pension schemes now will draw their pension in the 2050s, and they should expect their contributions to be invested in part to ensure that the world is in a better place than it would be if business as usual continued in the intervening period. There is the pragmatic challenge for a local green community that cuts through all the analysis in Whitehall! If the numbers do not add up, then it will be a matter of further public discourse until they do!

ACKNOWLEDGEMENT

The views and arguments are my own and do not represent official policy. The data have been corralled from many sources, and I am grateful to all those with whom I interacted during my period as Chief Scientific Adviser at DCLG 2006–9.

Index